Mohamed

I0019705

Coverage Strategies in Wireless Sensor Networks

Mohamed Khalil Watfa

Coverage Strategies in Wireless Sensor Networks

A rigorous analysis

VDM Verlag Dr. Müller

Impressum/Imprint (nur für Deutschland/ only for Germany)

Bibliografische Information der Deutschen Nationalbibliothek: Die Deutsche Nationalbibliothek verzeichnet diese Publikation in der Deutschen Nationalbibliografie; detaillierte bibliografische Daten sind im Internet über http://dnb.d-nb.de abrufbar.

Alle in diesem Buch genannten Marken und Produktnamen unterliegen warenzeichen-, marken- oder patentrechtlichem Schutz bzw. sind Warenzeichen oder eingetragene Warenzeichen der jeweiligen Inhaber. Die Wiedergabe von Marken, Produktnamen, Gebrauchsnamen, Handelsnamen, Warenbezeichnungen u.s.w. in diesem Werk berechtigt auch ohne besondere Kennzeichnung nicht zu der Annahme, dass solche Namen im Sinne der Warenzeichen- und Markenschutzgesetzgebung als frei zu betrachten wären und daher von jedermann benutzt werden dürften.

Coverbild: www.purestockx.com

Verlag: VDM Verlag Dr. Müller Aktiengesellschaft & Co. KG
Dudweiler Landstr. 99, 66123 Saarbrücken, Deutschland
Telefon +49 681 9100-698, Telefax +49 681 9100-988, Email: info@vdm-verlag.de
Zugl.: Norman, University of Oklahoma, Diss, 2006

Herstellung in Deutschland:
Schaltungsdienst Lange o.H.G., Berlin
Books on Demand GmbH, Norderstedt
Reha GmbH, Saarbrücken
Amazon Distribution GmbH, Leipzig
ISBN: 978-3-639-17335-2

Imprint (only for USA, GB)

Bibliographic information published by the Deutsche Nationalbibliothek: The Deutsche Nationalbibliothek lists this publication in the Deutsche Nationalbibliografie; detailed bibliographic data are available in the Internet at http://dnb.d-nb.de.

Any brand names and product names mentioned in this book are subject to trademark, brand or patent protection and are trademarks or registered trademarks of their respective holders. The use of brand names, product names, common names, trade names, product descriptions etc. even without a particular marking in this works is in no way to be construed to mean that such names may be regarded as unrestricted in respect of trademark and brand protection legislation and could thus be used by anyone.

Cover image: www.purestockx.com

Publisher:
VDM Verlag Dr. Müller Aktiengesellschaft & Co. KG
Dudweiler Landstr. 99, 66123 Saarbrücken, Germany
Phone +49 681 9100-698, Fax +49 681 9100-988, Email: info@vdm-publishing.com
Norman, University of Oklahoma, Diss, 2006

Copyright © 2009 by the author and VDM Verlag Dr. Müller Aktiengesellschaft & Co. KG and licensors
All rights reserved. Saarbrücken 2009

Printed in the U.S.A.
Printed in the U.K. by (see last page)
ISBN: 978-3-639-17335-2

COVERAGE STRATEGIES IN WIRELESS SENSOR NETWORKS

By

MOHAMED KHALIL WATFA, PhD

College of Computer Science and Engineering
University of Wollongong in Dubai (UOWD)
Email: Mohamed.Watfa11@gmail.com
Website: http://mohamedwatfa.synthasite.com/

To my family and loved ones…

I dedicate this book to my family and many friends. I would like to start by thanking God for his blessings and for sending me two angels from heaven, my beautiful wife, Diana, and my baby boy, Walid.

I also present a special feeling of gratitude to my loving parents, Khalil and Habiba whose words of encouragement and push for tenacity ring in my ears. My sister Lana, sisters in law Maya and Douja and brothers Walid and Ali have never left my side and are very special.

Acknowledgements

I would like to start by mentioning that this book is a direct result from my dissertation work between years 2003-2006. It was an incredible experience to pursue my Ph.D. at the University of Oklahoma in computer engineering. I learned a lot both from a professional as well as a personal angle, and I am still learning!

I would like to begin by thanking Prof. Sesh Commuri, who was my advisor and mentor for three years. He was a wonderful advisor, providing me support, encouragement, and always valuable advices during my Ph.D. experience. His enthusiasm for research and his vision for the future have been an inspiration. I thank him for the time he spent with me, discussing everything from research to career choices, and led me to this vibrant and challenging research field for wireless ad hoc and sensor networks. While at OU, I had the privilege of interacting with wonderful, bright, and talented people. I want to thank every single member of my advisory committee for taking the time to discuss my research ideas as well as their effort that was put in to completion of this dissertation. I would like to thank Dr. Sluss, Dr Refai, Dr Debrunner and Dr. Radhakrishnan.

I would like to acknowledge the continuous support of my beautiful and kind wife Diana who was there for me through every step and was always an umbrella that protected me from the rainy and harsh life. Diana, this book would not have been written without your patience and understanding.

Also, I would not be here without the support of my wonderful family. My father Khalil Watfa and my wonderful mom Habiba are the greatest parents any one would dream of having. Mom and Dad, I am who I am because of you and anything I would achieve in the near future is for you. Thank you for supporting me in every step and most importantly believing in me. To the two people I think guided me through every step, a thousand thanks to my brothers Walid and Allie. Walid you are an inspiration and a candle that lights up my path and gave me all the tools that I needed in order to succeed. You are a great brother and a friend as well. I wish you the best and I hope one day I could repay you and Allie all your good deeds.

   I am going to conclude my acknowledgments to all the people who were by my side with:

For all the ways you've cared for me,
For all the love you've shared with me,
For always being there for me,
Thank you!

Table of Contents

7.   Book Conclusion   216

List of Tables

List of Figures

Book Summary

A wireless sensor network is a network of a number of distributed nodes, each equipped with sensors, computational elements, and transceivers. These networks are able to sense desired phenomenon over a large geographic region and communicate this information back to the user or to a sink. Most of these networks are designed as stand alone networks comprising of thousands of low cost nodes for rapid deployment and are being used in a variety of applications from remote monitoring of habitats to military applications. While the use of these networks has been demonstrated, their full capabilities have not been realized, primarily due to the lack of efficient algorithms for self organization and fault tolerant operation.

A fundamental issue in the deployment of a large scale Wireless Sensor Network (WSN) is the ability of the network to cover the region of interest. While it is important to know if the region is covered by the deployed sensor nodes, it is of even greater importance to determine the minimum number of these deployed sensors that will still guarantee coverage of the region. This issue takes on added importance as the sensor nodes have limited battery power. Redundant sensors affect the communications between nodes and cause increased energy expenditure due to packet collisions. While scheduling the activity of the nodes and designing efficient communication protocols help alleviate this problem, the key to energy efficiency and longevity of the wireless sensor network is the design of efficient techniques to determine the minimum set of sensor nodes for coverage.

Currently available techniques in the literature address the problem of determining coverage by modeling the region of interest as a planar surface. Algorithms are then developed for determining point coverage, area coverage, and barrier coverage. The analysis in this book shows that modeling the region as a two dimensional surface is inadequate as most applications in the real world are in a three dimensional space. The extension of existing results to three dimensional regions is not a trivial task and results in inefficient deployments of the sensor networks. Further, the type of coverage desired is specific to the application and the algorithms developed must be able to address the selection of sensor nodes not only for the coverage, but also for covering the border of a region, detecting intrusion, patrolling a given border, or tracking a phenomenon in a given three dimensional space. These are very important issues facing the research community and the solution to these problems is of paramount importance to the future of wireless sensor networks.

In this book, the coverage problem in a three dimensional space is rigorously analyzed and the minimum number of sensor nodes and their placement for complete coverage is determined. Also, given a random distribution of sensor nodes, the problem of selecting a minimum subset of sensor nodes for complete coverage is addressed. A computationally efficient algorithm is developed and implemented in a distributed fashion. Numerical simulations show that the optimized sensor network has better energy efficiency compared to the standard random deployment of sensor nodes. It is demonstrated that the optimized WSN continues to offer better coverage of the region even when the sensor nodes start to fail over time. A localized 'self healing' algorithm is implemented that wakes up the inactive neighbors of a failing sensor node. Using the "flooding algorithm" for querying the network, it is shown that the optimized WSN with integrated self healing far outweighs the performance that is obtained by standard random deployment. For the first time, a 'measure of optimality' is defined that will enable the comparison of different implementations of a WSN from an energy efficiency stand point.

The analytical methods developed in this book are flexible and are shown to easily accommodate the requirements of different types of coverage encountered in the practical deployment of a wireless sensor network. First, given a region of interest, distributed algorithms are developed to select an optimal set of sensor nodes that cover the boundary of a region (Boundary Coverage Problem). Then, a distributed algorithm is also developed that enables the determination of the extent of coverage obtained by a WSN (Coverage Boundary Problem). In practice, several anomalies can occur in wireless sensor networks that impair their desired functionalities resulting in the formation of different kinds of holes such as coverage holes, routing holes, jamming holes, and worm holes. Determining the location and extent of the holes in the coverage is the first step in the augmentation of the network for improved performance. These issues are rigorously analyzed, and an algorithm that identifies the sensor nodes on the boundary of the coverage holes in the region is also developed.

Finally, a novel approach for tracking a dynamic phenomenon is presented. One of the central issues in sensor networks is energy efficient target tracking, where the goal is to monitor the path of a moving target using a minimum subset of sensor nodes while meeting the specified quality of service (QoS). Unlike other tracking methods that are based on computationally complex clustering techniques, the strategy adopted in this book is based on a computationally simple but elegant technique of finding a reduced cover of the whole region and then subdividing the reduced cover into sub-covers based on the target's location. The tradeoffs involved in target tracking are analyzed and the performance of the tracking algorithm is compared with other popular strategies from the literature. The behavior of the proposed tracking algorithm is analyzed through simulation and the improved performance is demonstrated. The gain in energy savings come at the expense of reduced quality of tracking. The algorithm presented guarantees the robustness and accuracy of tracking, as well as the extension of the overall system lifetime. The algorithms developed in this book are based on a number of reasonable assumptions that are easily verified in densely distributed sensor networks and require only a limited number of simple local computations.

The coverage algorithms developed in this book are a significant addition to the scientific knowledge in the area of wireless sensor networks. The proposed techniques help realize the practical deployment of wireless sensor networks in three dimensional regions. For the first time, the 'self healing' property along with the optimization techniques proposed herein make possible the implementation of highly efficient, robust sensor networks whose performance is optimized with respect to the needs of the application.

PART I

INTRODUCTION AND PROBLEM STATEMENT

Chapter 1

Introduction

A sensor network is a network of a number of distributed nodes, each equipped with sensors, computational elements, and transceivers. Sensor nodes form a sensing, computing, and communication infrastructure that allows us to instrument, observe, and respond to phenomena in the physical and cyber world. As sensor nodes are typically battery operated, it is important to efficiently use the limited energy of the nodes to extend the lifetime of the sensor network. Wireless Sensor Networks (WSNs) as they exist today, suffer from major disadvantages in their deployment and organization resulting in wastage of energy and thus reducing the overall system lifetime. Given an existing distribution of sensor nodes, it is often necessary to minimize the number of nodes that remain active while still achieving complete coverage of the entire region. If all the nodes are active simultaneously, an excessive amount of energy would be wasted due to packet collisions. Further, the data collected will also be highly correlated and redundant. The self organizing capacity of sensor networks for coverage of a three dimensional region and algorithms for energy efficient deployment of wireless sensor networks are fundamental issues facing the research community today and will be the focus of this book. This is a very challenging task as it involves several technological issues such as, coverage in three dimensional spaces, design of efficient networking protocols, power management, fault tolerance and accommodation, boundary coverage, intrusion detection, and distributed tracking algorithms. Successful solution to these research issues will promote the development of pervasive computing systems built using a network of tiny, wireless sensing nodes.

Aggregating sensor nodes into sophisticated sensing, computational and communication infrastructures to form wireless sensor networks will have a significant impact on a wide array of applications ranging from military, to scientific, to industrial, to health-care, to domestic, establishing ubiquitous computing that will pervade society redefining the way in which we live and work. Mark Weiser envisioned his view of *ubiquitous computing*, now also called *pervasive computing*, in his pioneer paper *"The Computer for the 21st Century"* [1] in the early 1990s. The essence of his vision was the creation of an environment saturated with computing and communication capability, yet gracefully integrated with human users [2]. In his paper, Mark Weiser wrote: *"The most profound technologies are those that disappear. They weave themselves into the fabric of everyday life until they are indistinguishable from it."*

Advances in hardware, software, and networking over the past decade have brought the vision of pervasive computing close to technical and economic viability [3]. The interest in distributed wireless sensor networks in academia and industry from the late 1990s is one of the most exciting and specific instance of ubiquitous computing effort. The advances in wireless communications and Micro-Electro Mechanical Systems

3

(MEMS) technologies have enabled the construction of a wide variety of wireless sensor/actuator devices that are small in size. These devices consist of one or more integrated sensing units, embedded microprocessors, low-power communication transceivers, and a small on-board power source. These devices also have location awareness and can be organized in an ad hoc multi-hop network. Besides sensor networks, the proliferation of inexpensive, widely available wireless devices and the network community's interest in mobile computing have fostered the rapid expansion of wireless ad hoc networks. Many future applications will increasingly depend on embedded wireless sensor and ad hoc networks, such as environmental monitoring, infrastructure maintenance, traffic management, energy management, disaster mitigation, personal medical monitoring, smart building, as well as military and defense. These broad application areas for wireless sensor and ad hoc networks will create a huge market in the foreseeable future. According to Business Communications Company, Inc., a market research firm, the U.S. market for industrial sensors will reach $7.6 billion by the year 2009 [4]. For this vision to become a reality, significant technological advances have to be made in the ability of these WSNs to self organize themselves to form intelligent networks capable of bridging the gap between the physical and the cyber world.

In this chapter, the motivations for the research presented in this book are first discussed. Then, a brief review of wireless networks, wireless ad hoc networks, and wireless sensor networks is presented in Section 1.2. The research challenges and the open issues in sensor and ad hoc networks is the focus of Section 1.3. The scope of this

4

book is presented in Section 1.4 and finally, the contributions of this book are summarized in Section 1.5.

1.1 Motivation

Advances in wireless communication and low-cost sensors have made possible the design and deployment of large-scale wireless sensor systems. Such networks are increasingly deployed in buildings, underwater, on roads or bridges, and in planetary exploration. Most existing results focus on planar networks [4]-[32]; however, three-dimensional modeling of the sensor network would more accurately reflect the real-life situations. 2D modeling of WSNs has been the focus of attention of most research works in the field of sensor networks due to its simplicity. However, modeling the coverage region as a planar surface results in inefficient implementations and therefore limits the use of WSN to a few applications. The potential application categories of sensor networks include:

1- *Military and Defense*: battlefield surveillance, reconnaissance of enemy forces and terrain, nuclear, biological and chemical attack detection and recon-naissance, etc.

2- *Environment*: bush fire monitoring, food detection, disaster mitigation, precision agriculture, etc.

3- *Health*: telemonitoring of human physiological data, tracking and monitoring doctors and patients inside a hospital, drug administration in hospital.

4- *Home*: home automation and networking, smart environment, etc.

5- *Other Commercial Areas*: environmental control in buildings, interactive museums, detecting and monitoring car thefts, managing inventory control, vehicle tracking and detection, etc.

Specifically, the following applications require the modeling of the coverage region of the WSN in the three dimensional space. The results presented in this book facilitate the efficient implementation of WSN to tackle these and similar problems.

1- *Disaster Recovery:* Natural disasters (floods, hurricanes, and fires) require sensing in different planes and thus 3-dimesnsional coverage techniques are required. Three-dimensional networks also arise in building networks where nodes are located on different floors.

2- *Mapping Topographical Properties:* Random dense sensor deployment on irregular terrains like mountains and hills leaves three dimensional coverage holes that indicate the topographical properties of the terrain. Understanding the topography of an area enables the understanding of watershed boundaries, drainage characteristics, water movement, impacts on water quality, and soil conservation.

3- *Space Exploration* [5], [6]: Wireless sensor networks will play an important role in planetary explorations. A rover functioning as a base station can collect measurement data from a number of mobile sensing units and relay the aggregated results to an orbiter.

4- *Undersea Monitoring* [7, 8 and 9]: Deployment of sensor nodes underwater enables the real time monitoring of selected ocean areas. Under Water Acoustic Sensor Networks (UW-ASN) can consist of a number of sensors and submersible vehicles that are deployed to perform collaborative monitoring tasks over a given area.

The applications mentioned above do not afford the flexibility in placing sensor nodes at desired locations for optimum coverage. As a practical matter, the sensor nodes are randomly distributed, for example dropped from an airplane, onto the region to be monitored. The sensing needs in each of these applications are fundamentally different and require the use of different types of sensors, each with a different sensing range and sensitivity. A sensor node typically has numerous sensors and can be configured to monitor the region and transmit the information back to a sink. The number of sensor nodes required depends on the size of the region to be covered, the sensing radius of the node, and the type of coverage desired. The information is transmitted to a sink typically through multi-hop communications in the WSN. The impact of the number of nodes on the capacity of multi-hop wireless networks was analyzed for deployments in two dimensions [10] and three dimensions [11]. Under a protocol model of non-interference, if n nodes, each with a transmission rate of W bits/second, are randomly distributed in a disc of area A sq. meters (m^2), then the throughput obtained by each node for transmission to a randomly chosen sink is given by $\Theta\left(\dfrac{W}{\sqrt{n\log n}}\right)$ $bits/\sec$ [10]. Similarly in the three dimensional deployment of wireless nodes, the throughput achieved when n nodes are located in a sphere of volume V is given by $\Theta\left(\dfrac{W}{\left(n\log^2 n\right)^{1/3}}\right)$ $bits/\sec$ [11]. Since the number of active nodes depends on the type of sensing required and the region of coverage, the overall communication and energy efficiency of the WSN can be significantly improved by optimizing the number of nodes while guaranteeing coverage of a region.

The above discussion shows that as a practical matter, the analysis of both the coverage of the WSN and the inter-node communications require modeling the network in a three dimensional space. The coverage problem is one of the fundamental issues in wireless sensor networks and due to the large number of deployed nodes and the high density of deployment, only a subset of the deployed nodes needs to be active in order to achieve the necessary quality of service. The 3D coverage problem in WSNs has not yet been addressed. Also, one of the important applications of WSNs is intrusion detection and tracking. Most of the work done in this area uses high computational clustering techniques in order to predict the target's next location and activate a subset of sensor

8

nodes accordingly. On the other hand, simple and energy efficient distributed algorithms that will allow a minimum subset of sensor nodes to be active in order to detect and track the intruder are provided in this book. The coverage problem was addressed in three dimensional space and distributed algorithms were provided in order to increase the system life time while achieving full coverage. The boundary coverage problem was also addressed where a reduced subset of sensor nodes were activated in order to detect an intruder to a region of interest at all times. Both results (full coverage, boundary coverage) were used to develop a novel approach to the energy efficient tracking problem using wireless sensor networks.

In the following sections, we will introduce wireless networks and discuss the challenges in wireless sensor networks.

1.2 Wireless Networks

Wireless networks can be broadly classified into two categories: *infrastructure-based* and *infrastructure-less* as shown in Figure 1.1 [12]. Infrastructure-based networks include traditional cellular networks and wireless LANs (with centralized control module). Infrastructure-less networks include ad hoc networks and sensor networks. Depending on the mobility of nodes, ad hoc networks can be further classified into mobile ad hoc networks (MANET) and static ad hoc networks.

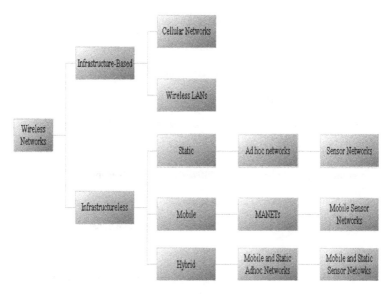

Figure 1.1: Classification of Wireless Networks.

1.2.1 Wireless Ad Hoc Networks

An ad hoc network is a collection of communication devices (nodes) that form a peer-to-peer network (no centralized server) temporarily, and in an ad hoc manner without any backbone infrastructure or base stations to meet immediate application needs. In ad hoc networks, individual nodes are responsible for dynamically discovering and selecting suitable communication neighbors to form a connected multi-hop network topology. A key assumption in ad hoc networks is that not all nodes can directly communicate with each other, so nodes are required to relay packets on behalf of other nodes in order to deliver data across the network [13]. Each node in an ad hoc network can operate as a client or server as well as a router. In a mobile ad hoc network, the network topology, connectivity, and node locations are variable and can be changed dynamically.

Ad hoc networks are one of the most vibrant and active research fields today. Significant research in this area has been ongoing for nearly 30 years. The history of ad hoc networks can be traced back to 1972 and the Department of Defense (DoD) sponsored *Packet Radio Networks* (PRNET), which evolved into the *Survivable Adaptive Radio Networks* (SURAN) program in the early 1980s [14]. The goal of these programs was to provide packet switched networking to mobile battlefield elements in an infrastructure-less, hostile environment (soldiers, tanks, aircrafts, etc., forming the nodes in the network) [15]. The interest in ad hoc networking was significant during the mid 1990s because of the advances of technologies, proliferation of laptop computers, and widespread use of wireless devices. The enabling technologies of ad hoc networks include the emergence of self-organizing systems, software defined radios (SDR), miniaturization of wireless devices, smart antenna, and battery technologies. The release of new frequency bands also provides the possibility of design and implementation of high speed data communication systems. Ad hoc networks are suitable for use in situations where infrastructure is either not available, not trusted, or is unavailable in times of emergency [13]. Some application examples include: battlefield communications; biological detection, environmental monitoring, target tracking; an infrastructure-less network of notebook computers in a conference or campus setting;

11

space explorations; undersea operations; etc. Although ad hoc networking research has a long history, significant challenges still exist. These challenges are summarized as follows.

1- *Scalability:* The dependence of the performance factors in an ad hoc network on the network size is addressed through scalability studies. Important among these is whether the ad hoc network can provide acceptable level of services (such as packet latency and network throughput) even when the number of nodes is large in the network. Scalability is especially important in sensor networks as sensors are normally deployed with large numbers to achieve large area coverage.

2- *Energy Efficiency:* Since ad hoc networks assume no fixed infrastructure, individual nodes have to rely on limited battery power. Energy efficiency therefore becomes an important issue in ad hoc networks. Low power operation is especially critical in sensor networks, as sensor networks are normally required to operate with a long lifetime. Recharging or replacement of batteries is either impractical or impossible for sensor networks after deployment.

3- *Quality-of-Service (QoS):* QoS is an active research area even in wired packet switching networks. Ad hoc networks further complicated the QoS challenges because of the error-prone and time-varying characteristics of wireless channels. Furthermore, individual nodes in ad hoc networks must share the media with many neighbors, each with its own set of QoS requirements.

4- *Security:* Security is another open issue for ad hoc networks since nodes normally use shared wireless media in a potential insecure environment. Nodes are susceptible to denial of service (DoS) attacks that are harder to track down than in wired networks.

5- *Lack of Well Defined System Models:* Finally, lack of well defined and widely accepted models for RF path attenuation, mobility, and traffic is another big issue. These tightly integrated models are required for a fair comparison and quantifying the system performance on a common baseline. Although the mechanisms behind electromagnetic wave propagation are well understood, it is difficult to quantify in detail in an environment including large number of complex objects (such as foliage, cars, and buildings).

1.2.2 Wireless Sensor Networks

Wireless sensor networks (WSNs) have become among the most vibrant and active research areas and have garnered much academia and industry attention. A simple sensor network is depicted in Figure 1.2.

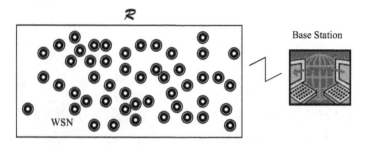

Figure 1.2: A Wireless Sensor Network (WSN) monitoring a region $\mathcal{R}$. By intelligently combining the data from the sensor nodes, the end user (base station) can remotely monitor events in the region of interest.

A sensor node is any device that maps a physical quantity from the environment to a quantitative measurement. Advances in sensor technology, low power analog and digital electronics, and low power radio frequency RF design have enabled the development of small, relatively inexpensive, low power sensors called micro sensors. Micro sensors are equipped with a sensor module e.g. acoustic, seismic, image sensor capable of sensing some quantity about the environment, a digital processor for processing the signals from the sensor and performing network protocol functions a radio module for communication and a battery to provide energy for operation. Each sensor node obtains a certain view of the environment as shown in Figure 1.2. A given sensors' view of the environment is limited both in range and in accuracy! It can only cover a limited physical area of the environment and depending on the quality of the hardware

14

may produce noisy data. Combining or aggregating the views of the individual nodes allows users to accurately and reliably monitor an environment.

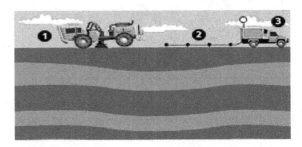

Figure 1.3: Modern-day oil prospectors use sound waves to locate oil. In one technique, (1) a signal is sent into the rock by a vibrator truck, (2) the reflected waves are received by geophones, and (3) the data is transmitted to a laboratory truck.

Wireless micro sensor networks represent a new paradigm for extracting data from the environment. Conventional systems use large, expensive macro sensors that are often wired directly to an end user and need to be accurately placed to obtain the data. For example, the oil industry uses large arrays of geophone sensors attached to huge cables to perform seismic exploration for oil as shown in Figure 1.3. These sensor nodes are very expensive and require large amounts of energy for operation. The sensors must be placed in exact locations since there are a limited number of nodes extracting information from the environment. Furthermore, deployment of these nodes and cables is costly and awkward requiring helicopters to transport the system and bulldozers to ensure the sensors can be placed in exact positions. There would be large economic and environmental gains if these large bulky expensive macro sensor nodes could be replaced with hundreds of cheap microsensor nodes that can be easily deployed. This would save significant costs in the nodes themselves, as well as in the deployment of these nodes. These sensor networks could be fault tolerant as the sheer number of nodes can ensure that there is enough redundancy in data acquisition even if all the nodes are not functional. Using wireless communication between the nodes would help eliminate the need for a fixed infrastructure.

A sensor network normally consists of a large number of sensor/actuator devices. Sensors nodes are normally powered by batteries and communicate untethered with short distances. Sensor networks represent a significant improvement over the traditional sensors, which are deployed in the following two ways [16]:

- Large, complex sensors can be positioned far from the phenomenon to be sensed. In this approach, complex signal processing algorithms and techniques may be employed to separate the target data from the environmental noise.
- Sensor nodes that perform only sensing tasks can be deployed. The positions of the sensors and network topology are carefully engineered. Individual sensors do not possess computational capabilities. They transmit the sensed phenomenon to one or more central nodes where data reduction and filtering can be performed.

Unlike traditional sensor deployment, future wireless sensor networks may be densely deployed either inside the phenomenon or in close proximity to it. The position of individual nodes need not be engineered or pre-determined. This allows random deployment of sensor networks in hostile or inaccessible terrains. On the other hand, random deployment requires that algorithms and protocols designed for sensor networks must possess self-organizing capabilities.

The structure of a sensor node can be seen from Figure 1.4. A sensor node is normally made up of four basic components: *a sensing unit, a processing unit, a transceiver unit,* and *a power unit.* A sensor node may also be equipped with *a location finding system, a mobilizer,* and *a power generator* dependent on applications. The hardware components are summarized as follows:

a) Processing Unit: Associates with small storage unit (tens of kilo bytes order) and manages the procedures to collaborate with other nodes to carry out the assigned sensing task.

b) Transceiver Unit: Connects the node to the network via various possible transmission media such as infrared, optical, radio and so on.

c) Power Unit: Supplies power to the system by small size batteries. This makes the onboard energy a scarce resource.

d) Sensing Units: It is usually composed of two subunits: sensors and analog-to-digital Converters (ADCs). The analog signal produced by the sensors are converted to digital signals by the ADC, and fed into the processing unit.

e) Other Application Dependent Components: Location finding system is needed to determine the location of sensor nodes with high accuracy; mobilizer may be needed to move sensor nodes when it is required to carry out the task.

On the other hand, the software platform of a sensor node consists of:

a) Embedded Operating System (EOS): Manages the hardware capability efficiently as well as supports concurrency-intense operations. Apart from traditional OS tasks such as processor, memory and I/O management, it must be able to rapidly respond to the hardware triggered events in real-time.

b) Application Programming Interface (API): A series of functions provided by OS and other system-level components for assisting developers to build applications.

c) Device Drivers: A series of routines that determine how the upper layer entities communicate with the peripheral devices.

d) Hardware Abstract Layer (HAL): Intermediate layer between the hardware and the OS. Provides uniform interfaces to the upper layer while its implementation is highly dependent on the lower layer hardware. With the use of HAL, the OS and the applications easily transplant from one hardware platform to another

The software platform of a typical sensor node is also shown in Figure 1.4(b).

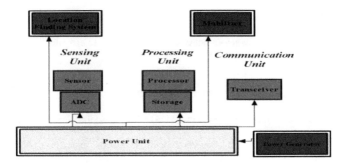

Figure 1.4 (a): A sensor node's hardware platform.

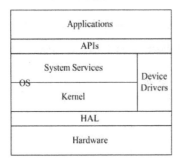

Figure 1.4 (b): A sensor node's software platform.

Sensor networks can be seen as a special case of ad hoc networks. Therefore, sensor networks suffer from the same drawbacks and challenges faced by ad hoc networks as discussed in the previous section. Sensor networks share some common features with ad hoc networks, such as infrastructure-less architecture and normally random network topology. However, sensor networks also have some different characteristics and more rigorous constraints compared to the broadly defined ad hoc networks. For instance, unlike nodes in an ad hoc network, sensors nodes are equipped with sensing units. Nodes in sensor networks are normally static while nodes in mobile ad hoc networks may change their locations rapidly. Also, the ability to interface with a wide variety of sensors makes sensor networks different from traditional ad-hoc networks.

In addition to these differences, sensor networks also possess the following characteristics which distinguish them from ad hoc networks and traditional wired and wireless networks.

a) *Resource Limitation*: Sensors are normally tiny devices and have stringent limitations for on-board energy, computational capability, and memory space. Energy consumption is the most important criterion to achieve long-life in sensor networks.

b) *Location Awareness*: Sensing data without knowing the location of the sensor is meaningless. Therefore, localization should be considered as an implicit feature of sensor networks.

c) *Data Centric*: Sensor networks are information driven and organized around the name of *data* instead of nodes. Applications express a need for a particular data element or type of data by naming it directly. For example, data query in sensor networks are typically in the style of "What is the temperature at location XYZ?" rather than "Connect node *A* to node *B*".

d) *Scalability*: The number of nodes in sensor networks may be several orders of magnitude higher than in ad hoc networks. Therefore, scalability is an important performance measurement for the protocols and algorithms designed for sensor networks.

e) *Dense Deployment*: Sensors are normally densely deployed with certain level of redundancy to achieve high sensing accuracy and collaborative data processing.

f) *Prone to Failures*: Sensors are prone to failure due to depletion of energy or physical damage.

These features of sensor networks introduce significant challenges that are different from traditional wired and wireless networks. The performance matrix for sensor networks can be classified as follows:

1- *Energy Efficiency/System Lifetime*: As sensors are normally battery powered and recharging or replacing the battery is impractical if not impossible, the protocol design must be energy efficient so as to maximize not only the lifetime of the individual nodes but also the lifetime of the entire network.

2- *Low Latency*: The observer is interested in knowing about the phenomena within a given delay and out-of-date information is of no use. Real-time guarantee is a pre-requisite for most sensor network applications.

3- *Sensing Accuracy*: Obtaining accurate sensing data is the primary objective of the observer, where sensing accuracy is an application dependent factor. The system efficiency can be further improved if network organization is based on sensing needs.

4- *Fault Tolerance*: Since sensors are prone to failures, sensor networks must be fault-tolerant so that non-catastrophic failures are hidden from the applications.

5- *Scalability*: Sensor network applications often feature a large number of sensor nodes. Therefore, the protocol design must be scalable.

Thus, it can be seen that while sensor networks are similar to ad hoc networks, the successful implementation of this technology should first address key issues in their deployment and performance. The open research issues will now be discussed in the next section.

1.3 Research Issues in Wireless Sensor Networks

In the last section, I briefly reviewed the wireless sensor networks and ad hoc networks. We also reviewed the research challenges and open issues. In this section, the goals of the research presented in this book will be identified. We will use the term *sensor* and *sensor node* inter-changeably in the rest of this book.

The main contribution of this book is the design and evaluation of a novel three dimensional coverage protocol for wireless sensor networks. In this book, contrary to existing techniques, the coverage problem in a three dimensional space is rigorously analyzed. The problem of determining the minimum number of sensors that guarantee complete coverage is first studied and an algorithm to choose a subset of working nodes for full coverage is derived. The cases of static and dynamic phenomena are also analyzed. The analysis is extended to handle the case of sensing and tracking a dynamic phenomenon in 3D.

The research issues related to wireless sensor networks are identified in Figure 1.5 and the contributions of this book are highlighted.

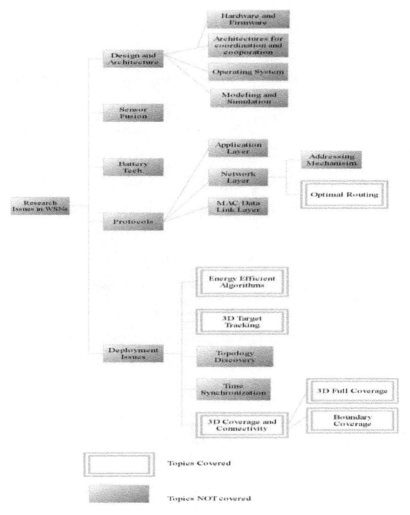

Figure 1.5: Wireless Sensor Network Research issues categorized.

24

1.4 Scope of the Book

As was presented in Figure 1.5, the following issues are addressed in this book:

- *Problem 1(Optimal 3D Deployment Problem):* Given a region **R**, what is the minimum number of sensor nodes that are needed to completely cover **R**?

- *Problem 2(3D Coverage Problem)*: Given a dense deployment of sensor nodes, find a minimum subset of active nodes that guarantee full coverage of the required region **R**.

- *Problem 3(Boundary Coverage)*: Given a dense deployment of sensor nodes, find a subset of active nodes that will guarantee the coverage of the boundary of a region of interest.

- *Problem 4(Coverage Boundary)*: Given a dense deployment of sensor nodes, find a subset of active nodes that lie on the boundary of the sensing coverage.

- *Problem 5(Tracking Problem)*: Given a dense deployment of sensor nodes, find a minimum subset of sensor nodes needed in order to track a dynamic phenomenon.

Algorithms for 3D coverage and connectivity will be derived when the phenomenon to be monitored is static. On the other hand, when the phenomenon is dynamic, algorithms for 3D tracking and connectivity will also be analyzed and studied.

1.5 Contributions

Energy efficient coverage of a region using Wireless Sensor Networks (WSNs) was addressed in this book. The main contribution of this book is a technique for obtaining a 3D reduced cover of a wireless sensor network. The technique is shown to be computationally simple and suitable for distributed implementation. Numerical simulations show that the reduced sensor network has better energy efficiency compared to the random deployment of sensor nodes. It was demonstrated that the reduced WSN continues to offer better coverage of the region even when the sensor nodes start to fail over time. A localized 'self healing' algorithm is implemented that wakes up the inactive neighbors of a failing sensor node. Using the "flooding algorithm" for querying the network, it is shown that the reduced cover of the WSN with integrated self healing offers superior performance over time. For the first time, a 'measure of optimality' has been defined that enables the comparison of different implementations of a WSN from an energy efficiency stand point.

The proposed algorithm is computationally simple and will result in lower communication overhead. The 3D coverage algorithm can be easily extended to obtain application specific reduced cover, border coverage for intrusion detection, to determine the mobility of sensor nodes to cover sensing holes, and to incorporate self-healing in sensor networks. Practical ways of 3D deployment for tracking applications is also one of the significant results of this book.

The theoretical and experimental contributions of this book will put the vision of deployment and self organization of WSNs closer to reality. The assumptions in this book are logical simplifications of the complex problem and result in algorithms that are elegant, efficient, and easy to implement in a distributed framework. Also, for the first time, an optimality measure is provided that compares one deployment strategy to another. In practice, sensor networks also introduce new challenges for fault-tolerance. Sensor networks are inherently fault-prone due to the shared wireless communication medium: message losses and corruptions (due to fading, collision, and hidden-node effect) are the norm rather than the exception. Moreover, node failures (due to crash and energy depletion) are commonplace. Since on-site maintenance is not feasible, sensor network applications should be self-healing. All the algorithms presented include self healing extensions, thereby increasing the robustness of the sensor network.

1.6 Book Structure

This book is organized into five parts. The first part introduces the problem and the motivation behind the research work presented in this book. The second part describes the 3D coverage problem, the first important problem of this book, and provides theoretical as well as experimental results. The third part deals with the boundary coverage problem and distributed algorithms for tracking using WSNs. The last part concludes the book.

Part I includes this chapter and Chapter 2. In Chapter 2, background and related work is surveyed.

Part II includes two chapters (Chapter 3 and Chapter 4) which provide detailed description and simulation results of the proposed coverage protocols for sensor networks. In Chapter 3, the motivations behind the design are first identified and the 3D coverage problem is formulated. We then propose a self healing algorithm as an extension to the coverage algorithm proposed in Chapter 4.

Part III includes two chapters (Chapters 5 and 6). Chapter 5 introduces the border coverage problem in wireless sensor networks and provides energy efficient algorithms. It also describes the coverage boundary problem and provides algorithms to select the optimal set of sensor nodes that lie on the boundary of coverage. Chapter 6 provides energy efficient target tracking algorithms using WSNs and compares the algorithm to other popular tracking algorithms and shows the results through simulations.

Part IV includes one chapter (Chapter 7). Chapter 7 concludes this book and identifies a number of remaining challenges and future research directions.

Chapter 2

Background and Related Work

In the previous chapter, wireless sensor networks were introduced and the motivations for the research presented in this book were highlighted. In this chapter, the background and literature relevant to this book is discussed. Topics covered include: coverage, connectivity issues and routing protocols and tracking using WSNs.

The rest of the chapter is organized as follows. In Section 2.1, related research works in coverage are summarized. The most discussed coverage problems from the literature can be classified in the following types: area coverage (Section 2.1.1), point coverage (Section 2.1.2), and barrier coverage (Section 2.1.3). The application of other types of coverage strategies is discussed in Section 2.1.4. The related research work in routing is then summarized in Section 2.2. Related work in routing in WSNs can be divided into flat-based routing (Section 2.2.1), hierarchical-based routing (Section 2.2.2), and adaptive based routing (Section 2.2.3) depending on the network structure. The chapter concludes with a discussion on related research work in tracking in Section 2.3.

2.1 Related Work in Coverage

An important problem addressed in literature is the **sensor coverage problem**. This problem is centered on a fundamental question: *"How well do the sensors observe the physical space?"* The research work related to the coverage problem are identified and summarized next.

As pointed out in [17], the coverage concept is a measure of the quality of service (QoS) of the sensing function and is subject to a wide range of interpretations due to a large variety of sensors and applications. The goal is to have each location in the physical space of interest within the sensing range of at least one sensor. The coverage algorithms proposed are either **centralized**, or **distributed**. In distributed algorithms, the decision process is decentralized. By distributed and localized algorithms, we refer to a distributed decision process that at each node that makes use of only neighborhood information (within a constant number of hops). Because the WSN has a dynamic topology and needs to accommodate a large number of sensors, the algorithms and protocols designed should be distributed and localized, in order to better accommodate a scalable architecture.

We survey recent contributions addressing coverage problems in the context of static WSNs, that is, the sensor nodes do not move once they are deployed. We present various coverage formulations, their assumptions, as well as an overview of the solutions proposed. The most discussed problems from the literature can be classified in the following types: area coverage, point coverage, and barrier coverage.

2.1.1 Area Coverage

The most studied coverage problem is the area coverage problem, where the main objective of the sensor network is to cover (monitor) an area (also referred sometimes as region). The area coverage could be divided into the following:

2.1.1.1 Random Area Coverage

Mechanisms that conserve energy resources are highly desirable, as they have a direct impact on network lifetime. Network lifetime is in general defined as the time interval that the network can perform the sensing functions and transmit data to the sink. During the network lifetime, some nodes may become unavailable (e.g. physical damage, lack of power resources) or additional nodes might be deployed. An efficient, frequently used mechanism is scheduling the sensor node activity and allowing redundant nodes to enter the *sleep* mode as often and for as long as possible. To design such a mechanism, the following questions should be addressed:

1. What is the rule that each node should follow to determine whether to enter sleep mode?

2. When should nodes make such a decision?

3. How long should a sensor remain in the sleep mode?

The scheduling of 'sleep/wake' activity in sensor nodes in a large distribution of randomly deployed nodes was first considered by [18] and [19] where they consider a large population of sensors, deployed randomly for area monitoring. Here, the goal is to achieve an energy-efficient design that maintains area coverage. This was achieved by first dividing the sensor nodes into disjoint sets where each set was capable of individually performing the area monitoring tasks. These sets are then activated successively, and while the current sensor set is active, the nodes in the remaining sets are in a low-energy sleep mode. For a given deployment, minimizing the number of active sensors is equivalent to finding the maximum number of disjoint sets. Minimizing the number of active sensors reduces the overall energy consumption and has a direct impact on the network lifetime. S. Slijepcevic and M. Potkonjak [18] have proved that the SET K-COVER which basically asks whether a set of sensors contain k disjoint sets to cover an area of interest is NP-complete problem using polynomial time transformation from the minimum cover problem. They developed the most-constrained least-constraining heuristic and demonstrated the effectiveness on variety of simulated scenarios. The theoretical analysis of the heuristic algorithm indicates that the worst runtime is $O(N^2)$, where N is the number of deployed sensor nodes. The basic idea of the approach is to minimize the coverage of sparsely covered areas within one cover. Such areas are identified using the notion of the *critical element*. The solutions proposed are centralized which is a major draw back and thus can not be used efficiently in a distributed sensor network.

Another energy-efficient node-scheduling-based coverage mechanism is discussed in [20]. D. Tian and N. D. Georganas [20] propose a node-scheduling scheme, which can reduce system overall energy consumption, therefore increasing system lifetime, by turning off some redundant nodes. Their coverage-based 'off duty eligibility rule' and 'backoff-based node-scheduling scheme' guarantees that the original sensing coverage is maintained after turning off redundant nodes. Their approach ensures that investigating whether the neighbors can cover the current node's sensing area is equivalent to checking whether the union of sponsored sectors (called sponsored coverage) contains the current node's sensing area, which in turn, is equivalent to calculating whether the union of central angles can cover the whole 360°. A probing-based, node-scheduling solution for the energy-efficient coverage problem is proposed in [21]. Here, all sensors are characterized by the same sensing range and coverage is seen as the ratio between the area under monitoring and total size of the network field. The off-duty eligibility rule is based on a probing mechanism. This protocol is distributed, localized, and has low complexity but still does not preserve the original coverage area.

Another work that is significant in the field of area coverage is [22]. Kumar et al. adopt the Randomized Independent Scheduling (RIS) mechanism in [22]. RIS assumes that time is divided into cycles based on a time synchronization method. At the beginning of each cycle, each sensor independently decides whether to become active with probability 'p' or go to sleep with probability '1 − p'. Thus, the network lifetime is increased by a factor close to $1/p$ (i.e. p determines the network lifetime). Kumar et al. derived the conditions for asymptotic K-coverage when RIS is used with three different deployment strategies – grid, random uniform and 2-dimensional Poisson. Assuming that active sensors are self-elected with a probability p using the Random Independent Scheduling (RIS) mechanism, Kumar et al. derived the sufficient conditions for achieving asymptotic k-coverage. Three deployment strategies were studied: grid ($\sqrt{n} \times \sqrt{n}$), random uniform (n points), and 2-dimensional Poisson (with rate n). They proved that for 'n' sensors deployed uniformly over a unit square region and assuming that active sensors are self-elected with a probability p using the Random Independent Scheduling (RIS) mechanism and some slowly growing function $\phi(np)$, if $c(n) = 1 + \dfrac{\phi(np) + k \log \log(np)}{\log(np)}$ then the unit square region is almost always k-covered. Their results are significant; however, they do not study the system lifetime and the resulting overhead in the communication.

In [23], C. Liu et al. deal with a challenging task which is determining how to schedule sensor nodes to save energy and meet both constraints of sensing coverage and network connectivity without accurate location information. Their approach utilizes an integrated method that provides statistical sensing coverage and guaranteed network connectivity. They use random scheduling for sensing coverage and then turn on extra sensor nodes, if necessary, for network connectivity. For a given point p in the field, they define the coverage intensity for this point as $C_p = \dfrac{T_c}{T_a}$ where T_a is any given long time period and T_c is the total time during T_a when point p is covered by at least one active sensor and the network coverage intensity, C_n, was defined as the expectation of C_p.

$C_n = E[C_p]$. They showed that without considering network connectivity, $C_n = 1 - (1 - \frac{q}{k})^n$, where q is the probability that each sensor node covers a given point.

It is very useful to dynamically adjust the coverage of a sensor network after it is deployed. When the total number of sensor nodes is fixed, the network coverage intensity can be adjusted by changing the number of disjoint subsets 'k'. The authors provided upper and lower bounds on the number of disjoint subsets D to provide a network

coverage intensity of at least t: $\left\lceil \dfrac{ln(1-t)}{ln(1-\frac{q}{k})} \right\rceil \leq D \leq \dfrac{q}{1 - e^{ln(\frac{1-t}{n})}}$. Their work is very useful,

as it provided bounds on the number of disjoint sets to achieve specific coverage intensity; however, their work does not indicate which sensor nodes and their locations are needed to be active after a random deployment. Also, they do no consider the system overhead and the resulting coverage lifetime.

     All the works in the area coverage limit their modeling to two dimensional regions and an optimality measure was not established in order to compare one sensor deployment strategy to another.

2.1.1.2 Connected Area Coverage

An important issue in WSNs is connectivity. A network is connected if any active node can communicate with any other active node, possibly using intermediate nodes as relays.

Once the sensors are deployed, they organize into a network that must be connected so that the information collected by sensor nodes can be relayed back to data sinks or controllers. An important, frequently addressed objective is to determine a minimal number of working sensors required to maintain the initial coverage area as well as connectivity. Selecting a minimal set of working nodes reduces power consumption and prolongs network lifetime. Next, several connected coverage mechanisms from the literature are reviewed.

An important but intuitive result, proved by Zhang and Hou [24], states that complete coverage of a convex area implies connectivity of the working nodes if the communication range is at least twice the sensing range. If the communication range is set up too large, radio communication may be subject to excessive interference. Therefore, if the communication range can be adjusted, a good approach to assure connectivity is to set transmission range as twice the sensing range. Wang et al., [25] generalized the result in [24] by showing that, when the communication range is at least twice the sensing range, a k-covered network will result in a k-connected network. A k-connected network has the property that removing any k-1 nodes will still maintain the network connectivity. The work in [25] introduces coverage configuration protocol (CCP) that can dynamically configure the network to provide different coverage degrees requested by applications.

2.1.1.3 Deterministic Area Coverage

In the last two sections, the sensor nodes are randomly deployed and 2D distributed algorithms are established for complete coverage. In this section, the deterministic placement of the sensor nodes is studied for coverage.

Kar and Banerjee [26] and references thereafter consider the problem of deterministically placing a minimum number of sensor nodes to cover a given region. K. Kar and S. Banerjee [26] address the problem of optimal node placement for ensuring connected coverage in sensor networks. They consider two different practical scenarios. In the first scenario, a certain region (or a set of regions) are to be provided connected coverage, while in the second case, a given set of n points are to be covered and connected. They first study the case in which the region to be covered is the entire two-dimensional plane. The solution for this simple special case provides some valuable insights for approaching the more complex region-coverage problems. They then discuss how regions of finite sizes can be provided connected coverage. Modeling each sensor node's sensing region as a disk or radius r, the authors proved that:

$$\frac{d_{P_1}}{d_{opt}} \leq \frac{\frac{\pi}{3} + \frac{\sqrt{3}}{2}}{1 + \frac{\sqrt{3}}{2}} = 1.026$$ where d_{P_1} denotes the density of the disk placement pattern P_1,

and d_{opt} denote the optimal density. They also consider the problem of providing connected coverage to a set of n given points in a two-dimensional Euclidean plane and show that $\frac{n_{p_3}}{n_{opt}} \le \frac{4\pi}{\sqrt{3}}$ where n_{p_3} denotes the number of disks used by the algorithm to provide connected coverage to the n given points and n_{opt} is the optimal solution. Their results are useful if we had the luxury in placing each sensor node in its particular location however that is not the case in most sensor networks. Also, all their results are for two dimensional sensor networks which are not too practical in real life.

2.1.1.4 Node Coverage as Approximation

When a large and dense sensor network is randomly deployed for area monitoring, the area coverage can be approximated by the coverage of the sensor locations. One method to assure coverage and connectivity is to design the set of active sensors as a connected dominating set (CDS). A distributed and localized protocol for constructing the CDS was proposed by Wu and Li, using the *marking process* in [27]. A node is a coverage node if there are two neighbors that are not connected (i.e., not within the transmission range of each other). Coverage nodes (also called gateway nodes) form a CDS.

In [28], Wu et al. studied the probabilistic conditions for complete redundancy, i.e. when a sensor's sensing area is completely covered by its neighbors' sensing areas. Then they studied the conditions for partial redundancy. They proved that given C, the sensing area of sensor S and its neighboring sensing areas Ci's ($1 \le i \le n$), if A is the event then C is fully covered by Ci's, then $1 - n0.609^{n-1} \le Pr\{A\} \le 1 - n0.609^{n-1} + \varepsilon$

where $\varepsilon = \dfrac{n(n-1)}{2}(0.276)^{n-1}$. They also showed that the average percentage of a

sensor's sensing area that is covered by its n random neighbors is not smaller

than $1 - n0.609^n - (\dfrac{n}{6} - 0.109)0.109^{n-1}$. Their results are important as it shows that it is

much more expensive to turn off nodes based on complete redundancy than partial

redundancy and based on these results, they proposed the Lightweight Deployment-

Aware Scheduling Mechanism (LDAS) to maintain statistical partial coverage.

2.1.2 Point Coverage

In the point coverage problem, the objective is to cover a set of points. Point coverage

research work is also divided into random point coverage and deterministic point

coverage each of which has different applications and approaches.

2.1.2.1 Random Point Coverage

The point coverage scenario addressed in [29] has military applicability. It considers a

limited number of points (targets) with known location that need to be monitored. A large

number of sensors are dispersed randomly in close proximity to the targets; the sensors

send the monitored information to a central processing node. The requirement is that

every target must be monitored at all times by at least one sensor, assuming that every

sensor is able to monitor all targets within its sensing range. One method for extending the sensor network lifetime through energy resource preservation is the division of the set of sensors into disjoint sets such that every set completely covers all targets. These disjoint sets are activated successively, such that at any moment in time only one set is active. As all targets are monitored by every sensor set, the goal of this approach is to determine a maximum number of disjoint sets, so that the time interval between two activations for any given sensor is the longest possible.

2.1.2.2 Deterministic Point Coverage

In [26], Kar and Banerjee consider the scenario where it is possible to explicitly place a set of sensor nodes. This is feasible in friendly and accessible environments. Given a set of n points, the objective is to determine a minimum number of sensor nodes and their location such that the given points are covered and all the deployed sensors are connected. For the case when all sensors have the same sensing range and the sensing range equals the communication range, the authors propose an approximation algorithm. The algorithm begins by constructing the minimum spanning tree over the targeted points, and then successively selects sensor node locations on the tree (vertices or along the edges) such that the coverage and connectivity is maintained at every step. The disadvantage of their approach is that constructing minimum spanning trees in a distributed fashion where each sensor node has only information about its communication neighbor is very difficult. Also, the computational complexity of forming such a spanning tree is another drawback.

2.1.3 Barrier Coverage

Inspired by Gage's classification [38], the barrier coverage can be considered as the coverage that minimizes the probability of undetected penetration through the barrier (sensor network). There are two types of barrier coverage models proposed in literature.

The first model is proposed by Meguerdichian et al. [30], where the following problem is addressed: given a field instrumented with sensors and the initial and final locations of an agent that needs to move through the field, determine a maximal breach path (MBP) and the maximal support path (MSP) of the agent. The MBP (MSP) corresponds to the worst (best) case coverage and has the property that for any point on the path, the distance to the closest sensor is maximized (minimized). The model assumes homogeneous sensor nodes, known sensor locations (e.g. through GPS), with sensing effectiveness decreasing as the distance increases. The authors proposed a centralized solution, based on the observation that MBP lies on the Voronoi diagram lines and MSP lies on Delaunay triangulation lines. The best coverage problem is further explored and formalized in [31], where Li et al. proposed a distributed algorithm for MSP computation using the relative neighborhood graph. Another important is the determination of the number of sensor nodes to be randomly deployed in the field such that the probability of a penetration path is close to zero. Liu and Towsley [32] address this coverage and detectability problem in the context of grid-based sensor networks and random sensor networks.

The second barrier coverage problem is the *exposure-based model*, introduced by Meguerdichian et al. in [33]. Exposure is directly related to coverage in that it is a measure of how well an object, moving on an arbitrary path, can be observed by the sensor network over a period of time. In addition to the informal definition, the authors formally define exposure and study its properties. They developed an efficient and effective algorithm for exposure calculation in sensor networks, specifically for finding minimal exposure paths. The minimal exposure path provides valuable information about the worst case exposure-based coverage in sensor networks. They express the general sensing model S at an arbitrary point p as $S(s,p) = \dfrac{\lambda}{[d(s,p)]^k}$ where $d(s,p)$ is the Euclidean distance between the sensor s and the point p, and positive constants λ and K are sensor technology-dependent parameters. Using this sensing model they define the exposure for an object O in the sensor filed during interval time $[t_1, t_2]$ along the path p(t) as: $E(p(t), t_1, t_2) = \int_{t_1}^{t_2} I(F, p(t)) \left| \dfrac{d(p(t))}{dt} \right| dt$ where $I(F, p(t)) = \sum_1^n S(s_i, p)$ is the sensor field intensity. The proved that the minimum exposure path from point p(1,0) to point q(0,1) is: $(\cos\frac{\pi}{2}t, \sin\frac{\pi}{2}t)$ and the exposure along this path is $E = \dfrac{\pi}{2}$. Their results are useful since they provided minimum exposure path for a given region however, their work limited the region of interest into a small 2D grid and they did not indicate the optimal location of the sensor nodes to maximize the exposure.

Another aspect of the exposure-based model is pointed out in [34]. To estimate the sensor node deployment density, one should consider both the sensor characteristics as well as target specifications. For example, detection of an enemy tank requires less nodes due to the strong the acoustic signal, compared with soldier detection that might require more sensors.

In the following section, some general coverage problems in other fields are introduced and this section is concluded.

2.1.4 Coverage Problem in Other Fields

Coverage problems have been formulated in other fields, such as the Art Gallery Problem and coverage in robotic systems. The Art Gallery Problem [35] is to determine the number of observers and their placement, necessary to cover an art gallery room such that every point is seen by at least one observer. This problem has a linear time solution for the 2D case. The 3D version is NP-hard and an approximation algorithm is presented in [36]. This problem has many real world applications, such as placement of antennas for cellular telephone companies, and placement of cameras for security purposes in banks and supermarkets. The work in [37] addresses the ocean area coverage problem. Here, the authors are interested in satellite based monitoring of the ocean phytoplankton abundance. Given the orbit and sensor characteristics of each mission, numerical analysis results show that merging data from three satellites can increase ocean coverage. The coverage concept with regard to the many-robot systems was introduced by Gage [38].

He defined three types of coverage: blanket coverage, barrier coverage, and sweep coverage. In the blanket coverage, the goal is to achieve a static arrangement of sensors that maximizes the total detection area. In barrier coverage the goal is to achieve a static arrangement of nodes that minimizes the probability of undetected penetration through the barrier, whereas the sweep coverage is more or less equivalent to a moving barrier.

Coverage is an important element for QoS in applications with WSNs. Coverage is in general associated with energy-efficiency and network connectivity, two important properties of a WSN. To accommodate a large WSN with limited resources and a dynamic topology, coverage control algorithms and protocols perform best if they are distributed and localized. Various interesting formulations for sensor coverage have been proposed recently in literature. None of which dealt with 3D sensor deployment. To meet the intended objective of the specific application, these problems aim at either deterministically placing sensor nodes, determining the sensor deployment density, or more generally, at designing mechanisms that efficiently organize or schedule the sensors after deployment. The coverage approaches and characteristics are summarized in Figure 2.1 and Table 2.1.

Figure 2.1: Coverage problems in WSNs divided into four different categories.

Table 2.1: Coverage Approaches in WSNs

Coverage Approach	Coverage Type	Problem Objectives	Deployment	Approach
Most / Minimal constrained	2D Area Coverage	Energy efficiency. Reduce # of working nodes.	Random	Centralized
Disjoint dominating sets	2D Area Coverage	Energy efficiency. Reduce # of working nodes.	Random	Centralized
Self Scheduling Algorithm	2D Area Coverage	Energy efficiency. Reduce # of working nodes.	Random	Distributed, Localized
Probing Based Density control	2D Area Coverage	Energy efficiency. Control working nodes density.	Random	Distributed, Localized
OGDC	2D Area Coverage	Energy efficiency. Connectivity. Reduce # of working nodes.	Random	Distributed, Localized
Coverage Configuration Protocol CCP	2D Area Coverage	Energy efficiency. Connectivity. Reduce # of working nodes.	Random	Distributed, Localized
Node placement Algorithms	2D Area Coverage Point Coverage	Energy efficiency. Connectivity. Deployment of minimum # of sensors	Deterministic	Distributed, Localized
Maximum Breach / Support Path	2D Barrier Coverage	Worst and best case coverage path	Random	Centralized
Minimum exposure path	2D Barrier Coverage	Find path of minimum exposure	Random	Centralized

46

In this book, contrary to existing techniques, the coverage and connectivity problem in 3D is rigorously analyzed. A distributed algorithm for establishing 3D coverage of a region of interest will be provided. We will also present an optimality measure that will allow you to compare one sensor deployment to another.

After we determined the coverage of the region of interest, the next step is to deliver sensed information from the sensor nodes, i.e., the sources, to the appropriate sinks. To do so, we need to design optimal routing algorithms. Works related to routing in WSNs are presented next. Any of the following routing algorithms could be used with the coverage algorithms presented in this book in order to deliver the necessary information to the sink.

2.2 Related Work in Routing

There may be several sinks in a sensor network. The sinks are gateways between the sensor network and the backbone network, e.g. Internet. Note that the sink may be in a ground-based site possibly set up by a rapid response team, in an unmanned airborne vehicle or plane, or a low earth orbit satellite. Depending on the required mission, the sinks and sensor nodes may be mobile. The objective of the *routing protocol* is to deliver sensed information from the sensor nodes, i.e., the sources, to the appropriate sinks. Sensed information will be represented by descriptors, which will be fused, i.e. if local neighbors have same descriptors, the descriptors will be combined, before they are routed to the sinks.

The routing protocol must meet the following design targets:

- *They must be power efficient.* Sensor nodes have low power capacity thus; power is a very important issue. The lifetime of a sensor node ends with the battery. As a result, redundant transmissions must be as low as possible.
- *They must be reliable.* Sensor nodes will deal with critical data in unreliable wireless environment.
- *Delays must be low.* The sensor network may also be used for real-time sensing. Thus, delay is an important issue.
- *Power emanation must be low.* In many missions the sensor network must be undetectable. Thus, the power emanation must be kept low.

In sensor networks, conservation of energy, which is directly related to network lifetime, is considered relatively more important than quality of data sent. As the energy gets depleted, the network may be required to reduce the quality of the results in order to reduce the energy dissipation in the nodes and hence lengthen the total network lifetime. Hence, energy-aware routing protocols are required to capture this requirement.

In general, routing in WSNs can be divided into *flat-based routing, hierarchical-based routing, and adaptive based routing* depending on the network structure. In flat-based routing, all nodes are assigned equal roles. In hierarchical-based routing, however, nodes will play different roles in the network. In adaptive routing, certain system parameters are controlled in order to adapt to the network current conditions and available energy levels. Furthermore, these protocols can be classified into multipath-based, query-based, negotiation-based, or location-based routing techniques depending on the protocol operation.

2.2.1 Flat Routing

Sequential Assignment Routing (SAR)

Routing decision in SAR [39] is dependent on three factors: energy resources, QoS on each path, and the priority level of each packet. To avoid single route failure, a multi-path approach is used and localized path restoration schemes are used.

Directed Diffusion:

Directed Diffusion [40] is a data-centric (DC) and application-aware paradigm in the sense that all data generated by sensor nodes is named by attribute-value pairs. The main idea of the DC paradigm is to combine the data coming from different sources in-network aggregation by eliminating redundancy, minimizing the number of transmissions; thus saving network energy and prolonging its lifetime.

Minimum Cost Forwarding Algorithm:

The minimum cost forwarding algorithm (MCFA) [41] exploits the fact that the direction of routing is always known, that is, towards the fixed external base-station. Each node maintains the least cost estimate from itself to the base-station.

Energy Aware Routing:

In [42], a destination initiated reactive protocol is proposed to increase the lifetime of the network. This protocol is similar to Directed Diffusion [42]. However, this protocol maintains a set of paths instead of maintaining or enforcing one optimal path.

2.2.2 Hierarchical Routing

LEACH protocol

Heinzelman, et al. [43] introduced a hierarchical clustering algorithm for sensor networks, called Low Energy Adaptive Clustering Hierarchy (LEACH). LEACH is a cluster-based protocol, which includes distributed cluster formation. LEACH randomly selects a few sensor nodes as cluster heads (CHs) and rotates this role to evenly distribute the energy load among the sensors in the network.

Threshold-sensitive Energy Efficient Protocols (TEEN and APTEEN):

Two hierarchical routing protocols called TEEN (Threshold-sensitive Energy Efficient sensor Network protocol), and APTEEN (Adaptive Periodic Threshold-sensitive Energy Efficient sensor Network protocol) are proposed in [45] and [46] respectively. These protocols were proposed for time-critical applications. In TEEN, sensor nodes sense the medium continuously, but the data transmission is done less frequently. A cluster head sensor sends its members a hard threshold, which is the threshold value of the sensed attribute and a soft threshold, which is a small

Virtual Grid Architecture routing

An energy-efficient routing paradigm is proposed in [47] that is based on the concept of data aggregation and in-network processing. The data aggregation is performed at two levels: local and then global.

Hierarchical Power-aware Routing

In [48], a hierarchical power-aware routing was proposed. The protocol divides the network into groups of sensors. Each group of sensors in geographic proximity is clustered together as a zone and each zone is treated as an entity. To perform routing, each zone is allowed to decide how it will route a message hierarchically across the other zones such that the battery lives of the nodes in the system are maximized.

2.2.3 Adaptive Routing

Heinzelman et al. in [49] and [50] proposed a family of adaptive protocols called Sensor Protocols for Information via Negotiation (SPIN) that disseminate all the information at each node to every node in the network assuming that all nodes in the network are potential base-stations. These protocols make use of the property that nodes in close proximity have similar data and thus distribute only the data that the other nodes do not have. All the routing techniques are summarized in Figure 2.2.

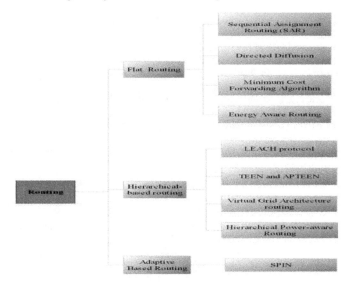

Figure 2.2: Routing Techniques in WSNs

Implementing optimal routing protocols is part of the future work. The routing scenario will be in a three-dimensional space and will also cover the case of routing to different sinks depending on the location of the phenomenon. Coverage problems related to static phenomenon have been surveyed. However, there is also the case where the phenomenon is dynamic. Works related to tracking a dynamic phenomenon is presented next.

2.3 Related Work in Tracking

In the previous two sections, related works in coverage and routing were summarized. In this section, one of the significant applications of sensor networks is highlighted and the research work in this area is discussed. One of the most important areas where the advantages of sensor networks can be exploited is for tracking mobile targets. Scenarios where such network may be deployed can be both military (tracking enemy vehicles, detecting illegal border crossings) and civilian (tracking the movement of wild animals in wildlife preserves). Typically, for accuracy, two or more sensors are simultaneously required for tracking a single target, leading to coordination issues. Additionally, given the requirements to minimize the power consumption due to communication or other factors, we would like to select the bare essential number of sensors dedicated for the task while all other sensors should preferably be in the hibernation or off state. In order to simultaneously satisfy the requirements like power saving and improving overall efficiency, we need large scale coordination and other management operations. These tasks become even more challenging when one considers the random mobility of the targets and the resulting need to coordinate the assignment of the sensors best suited for tracking the target as a function of time.

Target tracking is considered a canonical application for wireless sensor networks, and work in this area has been motivated in large part by DARPA programs such as SensIT. Zhao et al. present the information driven sensor querying (IDSQ) mechanism in [51], [52]. IDSQ is a sensor-to-sensor leader handoff based scheme in which at any given time there is a leader sensor node which makes the decisions about which sensors should be selectively turned on in order to obtain the best information about the target. A combined cost function which gives weight to both energy expenditure and information gain is considered. Liu et al. develop a dual-space approach to tracking targets which also enables selective activation of sensors based on which nodes the target is likely to approach next. Along these lines, Brooks et al. advocate a location-centric approach to performing collaborative sensing and target tracking in [53], [54]. The idea is to develop programming abstractions that provide addressing and communication between localized geographic regions within the network rather than individual nodes. This makes localized selective-activation strategies simpler to implement. They present self-organized distributed target tracking techniques with prediction based on Pheromones, Bayesian, and Extended Kalman Filter techniques [55], [56]. The implementation and testing of a real distributed sensor network collaborative

54

tracking algorithm in a military context is described in [57]. Important contributions by other groups on which we build include [58], [59] and [60]. K. Chakrabarty and S. Iyengar [59] present novel grid coverage strategies for effective surveillance and target location in distributed sensor networks. They represent the sensor field as a grid (two or three-dimensional) of points (coordinates) and use the term target location to refer to the problem of locating a target at a grid point at any instant in time. They first present an integer linear programming (ILP) solution for minimizing the cost of sensors for complete coverage of the sensor field. They solve the ILP model using a representative public-domain solver and present a divide-and conquer approach for solving large problem instances. They then use the framework of identifying codes to determine sensor placement for unique target location and provide coding-theoretic bounds on the number of sensors and present methods for determining their placement in the sensor field. They provided lower and upper bounds on S_n^p, the number of sensors required for uniquely identifying targets in an n-dimensional (n≤3) sensor field with p grid points in each dimension and is given as: $\frac{p^n}{n+1} \leq S_n^p \leq \frac{p^n}{n}$. Their work is important since it bounds the number of sensor nodes required to identify intruders; however they assumed that the sensors could be manually placed and they didn't provide a selection algorithm where a subset of already deployed sensor nodes can be selected to be active for target detection.

Massively distributed sensor networks are becoming a reality, largely due to the availability of the Mote hardware [61]. In [62], Cerpa and Estrin propose an adaptive self-configuring sensor network topology in which sensors can choose to join the network based on the network condition, the loss rate, the connectivity, etc. The sensors do not move, but the overall structure of the network adapts to the situation by having the sensors activate and deactivate. The addition of motion capability to the Mote sensors, creating Robomotes, was described in [63]. Algorithmic work has included even dispersal of sensors from a source point and redeployment for network rebuilding [64], [65]. Related recent work by Bullo et al. [66] uses Voronoi methods to arrange mobile sensors. Related research also includes mobile robotics work focused on distributed formation control [67-71]. Voronoi diagrams have been used in a variety of mobile robot research, but almost always it is the Voronoi diagram of the environment that is considered.

In this research, an architecture for managing and coordinating a sensor network for tracking 3D moving phenomena is proposed. We will deal with the case when the target to be tracked is big requiring a large number of sensors. We will also require the network to be connected at all time so that necessary information could be routed to the sink when needed. The tracking algorithm is specifically aimed at addressing the various challenges outlined while accurately tracking moving targets. The tracking algorithm should not require any central control point, eliminating the possibility of a single point of failure and making it robust against random node failures. The tracking task is carried out in a distributive manner by sequentially involving the sensors located along the track of the moving target.

In this chapter, the research work in the areas of coverage, routing and tracking were outlined and the pros and cons were highlighted. The gaps in the research work were discussed and the motivation behind the work was emphasized. In the next chapter, the first important problem addressed in this book is formulated, and distributed algorithms for its solution are presented. The performance of the algorithms is investigated through numerical examples.

PART II

THE THREE DIMENSIONAL
COVERAGE PROBLEM IN WSNs

Chapter 3

Three Dimensional Coverage

In the last chapter, related research work was surveyed and general approaches to the coverage problem were summarized. In this chapter, the 3D coverage problem is rigorously analyzed and energy efficient distributed algorithms are provided.

The energy efficiency of a WSN is studied in the context of the coverage of a region using wireless sensor nodes. First, the minimum number of sensor nodes required to cover a three-dimensional region is determined and an algorithm for testing the coverage of a sensor network is proposed. A computationally simple method for selecting a minimum subset of a random distribution of sensor nodes for coverage is then developed. This method is implemented in a distributed fashion across the WSN and the saving obtained is analyzed. Using the standard flooding algorithm [72]-[74], the performance of the optimized network is shown to be superior to the performance of the original deployment. A simple localized 'self healing' algorithm is also implemented that wakes up the inactive neighbors of a failing sensor node. It is shown that the performance of the optimized WSN with integrated self healing far outweighs the performance that is obtained by standard random deployment. For the first time, a 'measure of optimality' is defined that will enable the comparison of different implementations of a WSN from an energy efficiency stand point.

The rest of the chapter is organized as follows. The coverage problem in WSNs is formulated in Section 3.1. In Section 3.2, an algorithm for testing the coverage of a known deployment is presented. An optimal 3D deployment strategy is presented in Section 3.3. In Section 3.4, an algorithm for the selection of a minimum set of sensor nodes to guarantee coverage is developed. Another coverage selection algorithm is presented in Section 3.5. The connectivity and power efficiency issues are addressed in Section 3.6. Numerical simulation results that validate the proposed algorithms are presented in Section 3.7 and the chapter is concluded in Section 3.8.

3.1 The Coverage Problem Formulation

One of the fundamental problems in wireless sensor networks is determining how many sensor nodes are required to cover a specific area. For the purpose of the work in this book, each sensor node is assumed to be equipped with a radio interface and has the ability to communicate directly with other nodes in its vicinity. Since a sensor network consists of a large number of sensor nodes either placed at specific locations or distributed randomly in a region, it is advantageous to know ahead of time the dependence of the coverage on the deployment of the sensor nodes. Towards this end, the notion of sensing region and coverage are first defined and the coverage and optimization problems are then formulated. Since in practice, a large number of sensor nodes are distributed randomly in the region to be monitored, the sensing regions of all nodes are assumed to be identical. It is also assumed that the region to be monitored is large in comparison to the sensing region of an individual node and that the locations of all the nodes are known. The discussion in this chapter assumes the ability of each node to detect the quantity of interest. The actual sensed value, sensor fusion, and routing of the information to a sink, while of enormous importance, are beyond the scope of the present work.

3.1.1 Sensing Region

Let O be the output of a sensor node S that is capable of sensing a phenomenon P. Let S have a sensing radius R_s and a communication radius R_c.

Definition 3.1: The phenomenon P located at $Y \in R^3$ is said to be **detectable** by sensor S located at $X \in R^3$ if and only if there exists a constant threshold $\delta > 0$ such that $O(Y) > \delta$ if the phenomenon P is present. The quantity 'δ' is the signal threshold and is specific to the type of sensor used. □

The sensing region of sensor node S located at $X(x,y,z)$ is the collection of all points where the phenomenon P is detectable by the sensor node S, i.e. $A = \{ y \in R^3 \mid P \text{ is detectable by } S \}$. While the sensing region of an individual node can depend on the sensor and the environment it is deployed in, it is necessary to have a simplified representation of the sensing region to reduce the computational complexity in determining the coverage of a WSN. In this chapter, we will restrict the sensing region of S to be an <u>open ball</u> centered at $X \in R^3$.

<u>*Definition 3.2:*</u> Let $Y = \{ y \in R^3 \mid O(y) > \delta \}$. The **sensing region** of sensor node S located at $X \in R^3$ is defined as $A = \{ y \in Y \mid \|y - X\| \le R_s \}$, where $\|.\|$ is the Euclidean distance between y and X. $R_s \left(= \min_{Y_{(O(y) > \delta}} \|Y - X\| \right)$ in this definition is the radius of the open ball representing the sensing region and is called the sensing radius of S. ▢

Most of the research works thus far assume simplified Boolean sensing model (Circular disc) for coverage for protocol design and evaluation. In this model all events within the circular disc are assumed to be detected with probability 1. This simplified model is clearly not applicable to all types of sensing measurements however since the sensor nodes will be deployed in large numbers and there is a need for simulation and theoretical analysis, lead researchers in the areas of wireless sensor networks use this model in their research [14-26].

3.1.2 Problem Analysis

Consider a network comprising of sensor nodes $S_1, ..., S_n$, each with a sensing radius equal to R_s. Let A_i be the sensing region of node S_i and 'R' the region to be monitored. Then, given the collection of nodes $C_n = \{S_1, ..., S_n\}$, the sensing region of this collection can be expressed as $\overline{C_n} = \bigcup\limits_{i=1}^{n} A_i$.

Definition 3.3: A collection of sensor nodes $C_n = \{S_1, ..., S_n\}$ is said to cover the region R if and only if $R \subseteq \overline{C_n}$.   □

If C_n covers **R**, then $p \in \mathbf{R} \Rightarrow p \in A_i$ *for some* $S_i \in C_n$. This is commonly referred to in literature as the 1-cover of the region **R**. A natural extension of 1-cover is *k-cover* where any point in the region belongs to the sensing regions of at-least k nodes. In this chapter, cover is interpreted as the 1-cover of the region.

The coverage and deployment problem in WSNs can be addressed by studying the following three sub-problems:

- Problem 1(Coverage): Given a collection of sensor nodes $C_n = \{S_1, ..., S_n\}$ in a region **R**, determine if **R** is covered.

- Problem 2(Optimal Deployment): Find the minimum number of sensor nodes and their locations to cover a given region **R**.

- Problem 3(Reduced Coverage): Given a dense deployment of sensor nodes, find a reduced subset of active nodes that guarantee coverage of **R**.

3.2 Testing for Coverage

The objective of the coverage problem is to check if every point in the region of interest is covered by at least one sensor node in the network. In addition to determining the coverage, it is also advantageous to determine the degree of the cover and, in the case of incomplete coverage, the 'size' of the holes. The degree of the cover is the least number of sensor nodes that cover any point in the region and is a measure of redundancy in the WSN. Redundant active nodes are indicative of the excess energy expended in a WSN. The information on the size and location of the holes in the coverage helps determine the number of additional nodes required and their placement to guarantee complete coverage. The algorithm presented in this section tests for coverage by generating an occupancy grid and checking if each cell in this grid is covered by at least one node. If all the cells in the grid are occupied, then the entire region is covered. The procedure to generate this grid and the computational complexity are now discussed.

The coverage problem in this chapter is first posed as a problem of covering a 2D region with equal, overlapping disks. The region to be covered is divided into squares of side equaling half the radius of the disks representing the sensing region of a sensor node. The algorithm presented verifies if each cell in this grid is covered by at least one node. Since the region to be covered is divided into a grid with cell size equal to $\frac{R_s}{2}$, any cell in this grid is completely covered if its center is within a distance of $\frac{R_s}{2}$ from the sensor node. The factor '2' is chosen because it is the smallest integer that will satisfy this criterion. This criterion converts the problem of checking the coverage of a cell by a sensor node into the simpler problem of checking the distance between the center of the grid and the sensor node. As can be seen from Figure 3.1(a), at most 25 cells need to be checked to verify the coverage of a sensor node and a maximum of $25n$ cells need to be checked for the coverage region of 'n' nodes.

The coverage of a region is determined by first partitioning the region into a grid of spacing $\frac{R_s}{2}$. An origin is arbitrarily assigned and successive cells are numbered (Figure 3.1(b)). An occupancy matrix is then formed and initialized to zero. Every cell in the region is referenced by one entry in the occupancy matrix. For example, if the region is divided into 'p' rows and 'q' columns, then the ij-th cell is referenced by the $i*q+j$ entry in the occupancy matrix. Since each node can only cover a region of 5x5 cells, the cells covered are easily determined from the location of a given node. Thus for each node in the network, the covered cells, and thereby the entries in the occupancy matrix, are determined. The cells in the occupancy matrix corresponding to the covered cells are then indexed by '1'. A zero entry in the occupancy matrix indicates an uncovered cell in the region. The number of uncovered cells and their locations can then be used to determine the size and locations of the uncovered regions. Further, since the cell entries in the occupancy matrix are indexed, an entry 'k' indicates that the cell is covered by 'k' sensor nodes. Thus, the smallest entry in the occupancy matrix gives the degree of the cover.

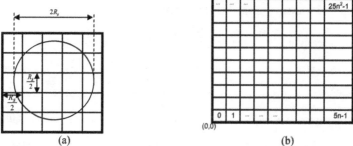

(a)   (b)

Figure 3.1: (a) The coverage region of a sensor (b) Coverage grid for 'n' sensors

The grid generated is the smallest grid entirely covering the region **R**. If **R** is smaller than the grid, then only relevant cells in the grid can be chosen for testing the coverage. This is done by assigning 'X', i.e. don't care, to cells that are outside the desired region of coverage. In the three dimensional case, the region is divided into cubes of side $\frac{R_s}{2}$ and coverage of each cell is verified by checking the distance between the node and the center of the cell.

It can be seen that this algorithm requires $O(n)$ steps to verify if a region is completely covered. The algorithm presented is simple and easy to implement. It not only determines the extent of coverage but also identifies the size and location of the holes in the coverage.

3.3 3D Deployment Strategy for the Sensor Nodes

The concepts of an optimum cover and reduced cover are introduced in this section. In a deployment of sensor nodes, if the number of the sensor nodes and their locations can be specified, then it is possible to find the exact location for each sensor node in order to cover the region using the smallest number of nodes. Such a deployment is **optimal**. On the other hand, given a random distribution of the sensor nodes, it is possible to activate only a subset of the nodes and still maintain cover of the region. A **reduced** cover is that set of sensor nodes that completely cover the region and removal of any node in this set leads to loss of cover. In this section, the problem of determining the optimum cover is addressed by solving the related problem of packing a volume with equal overlapping spheres. In order to achieve this, the notion of "thickness" of a cover is first defined along the lines in Conway [79].

Definition 3.3.1: The **thickness** 'θ' of a sensor cover is defined as the average number of sensor nodes that cover a point in the space.

θ = volume of one sensing region / volume of the fundamental region, i.e.

$$\theta = \frac{\frac{1}{n}\sum_{i=1}^{n}V_i}{\frac{1}{n}V_{total}} = \frac{nV}{V_{total}}, \text{ where 'n' is the number of active nodes; } V_i \text{ is the volume of sensing}$$

region of node s_i; and V_{total} is the total volume of the sensed region

R. *But since* $V_1 = V_2 = ... = V_n = V$, $\theta = \frac{nV}{V_{total}}$.   □

Definition 3.3.2: The **covering radius**, R_{cover}, of spheres centered at $X_1, X_2, ..., X_n$ is the minimum sensing radius that will cover the region **R**.   □

Intuitively, it can be seen that the thickness of a cover does not imply local efficiency of the deployment. For example, *1-cover* of a region **R** implies that there are points in the region that are covered by only one sensor node. However, there can be sensor nodes that can be removed without impacting the overall coverage. Thus, to determine the minimal deployment of the WSNs for coverage of a region, it is necessary to introduce the concept of a reduced cover.

Definition 3.3.3: A cover C_n of **R** with sensor nodes $S_1, S_2, ..., S_n$ each with sensing radius R_s and sensing regions $A_1, A_2, ..., A_n$ is **reduced** if no proper subset of C_n is a cover of **R**.   □

Then, the following statements are equivalent:

- A cover C_n of **R** is **reduced**.

- If a finite number of nodes of a cover C_n are removed, then C_n is no longer a cover of **R**.

Lemma 3.3.1: An optimum deployment of the nodes $C_n = \{S_1, S_2, ..., S_n\}$ with sensing radius R_s is reduced. □

Definition 3.3.3 and Lemma 3.3.1 together imply that an optimum deployment is reduced while the converse need not necessarily be true.

Theorem 3.3.1: Let **D** be the deployment of the sensor nodes $S_1, S_2, ..., S_n$ at the vertices '*L*' of a body centered cubic (bcc) lattice spanning a region $R \subset R^3$. Let Λ be the distance between adjacent vertices of L and R_s be the sensing radius of each sensor node. Then the deployment **D** is optimal if the lattice spacing $\Lambda = \dfrac{R_s}{1.118}$.

Proof: In the three-dimensional space, the thinnest cover of a region by spheres is obtained when the centers of the spheres are at the vertices of a body centered cubic (bcc) lattice [75, 76]. If the distance between adjacent vertices in this case is one unit, then the entire region can be covered by copies of a sphere whose covering radius is

$R_{cover} = \dfrac{\sqrt{5}}{2} = 1.1180$. Such a lattice is periodic and is completely reduced. Moreover, the

thickness of such a cover is $\theta = \dfrac{5\pi\sqrt{5}}{24} = 1.4635$.

Conversely, given a *bcc* lattice that spans a region of interest, this region can be

completely covered by spheres of radius R_s if the lattice spacing is $\Lambda = \dfrac{R_s}{R_{cover}} = \dfrac{R_s}{1.118}$.

Thus, the deployment **D** will be optimal if the spacing between the centers of adjacent

nodes equals $\dfrac{R_s}{1.118}$. $\qquad\qquad\qquad\qquad\qquad\qquad\qquad\qquad$ □

Theorem 3.3.1 implies that the arrangement of the sensor nodes with their centers coincided with the vertices of a *bcc* lattice will guarantee an optimal covering of the region spanned by this lattice. The spacing Λ of the lattice is a function of the sensing radius of the sensor node. The vertices of the lattice that lie within the sensing region **R** are the minimum number of nodes needed to cover the whole region **R**.

In a two-dimensional space, the thinnest covering of a region with circles is obtained when the centers of the circles lie at the vertices of a hexagonal lattice. If the distance between adjacent vertices in this case is one unit, then the entire region can be covered by copies of a disc whose covering radius is $R_{cover} = \dfrac{1}{\sqrt{3}} = 0.5773$. Such a lattice is also periodic and completely reduced. Moreover, the thickness of the cover is $\theta = \dfrac{2\pi}{3\sqrt{3}} = 1.2092$. Thus, the deployment $\mathbf{D}$ will be optimal if the spacing between the

centers of adjacent discs equal $\dfrac{R_s}{0.5773}$.

3.4 Efficient Sensor Selection for Complete Coverage in 3D

The algorithm in Section 3.3 enables the optimum placement of sensor nodes for complete coverage of a given region. In practice, however, given an existing distribution of sensor nodes, it is often necessary to minimize the number of nodes that remain active while still achieving complete coverage of the entire region. If all the nodes are active simultaneously, an excessive amount of energy would be wasted due to packet collisions. Further, the data collected will also be highly correlated and redundant. In this section, an algorithm is developed where the sensor nodes make local decisions on whether to sleep, or join the set of active nodes. If the sensing region of a node is completely covered by its neighbors, then the node can be disabled without affecting the overall coverage. Thus, by iteratively disabling nodes that are covered by other nodes, one can arrive at a reduced set of sensor nodes that guarantee a *1*-cover of the desired region. Since the initial distribution of the sensor nodes is random in nature, the nodes are not located at the vertices of a *bcc* lattice and therefore, the reduced set of nodes is not optimal. Thus, for the algorithm to be truly useful, it is necessary to also have a method to compare it to the optimal solution to ascertain the effectiveness of the proposed algorithm. In this section, a rigorous mathematical proof of the proposed technique is provided along with a metric to compare the performance of the reduced network with the optimum coverage obtained in Section 3.3.

Let $C_n = \{S_0, S_1, S_2, ..., S_n\}$ be the set of all the nodes that cover the region **R**. Also, let node S_0 have overlapping regions of coverage with sensors $S_1, S_2, ..., S_n$. Node S_0 can be deactivated if it is entirely covered by the nodes $S_1, S_2, ..., S_n$. Since it is impossible to verify if each and every point in the sensing region of S_0 is covered by some other node, a simplified technique is now proposed.

Let B_i be the boundary of the sensing region of node S_i, i.e. B_i is the surface of the ball representing the sensing region of S_i. Denote the portion of B_i inside A_0 as sur_i (See Figure 3.2(a)) and the intersection of two such surfaces as an arc, i.e. $arc_{ij} = sur_i \cap sur_j$ (See Figure 3.2(b))

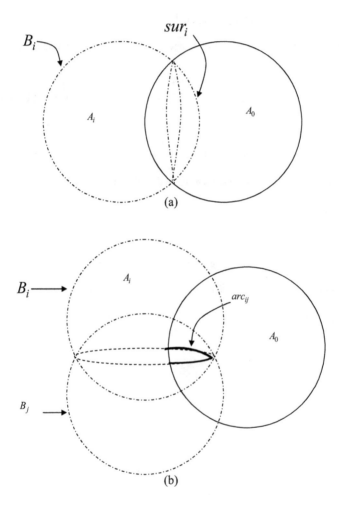

Figure 3.2: (a) $sur_i - B_i \cap A_0$ (b) $arc_{ij} - sur_i \cap sur_j$

Theorem 3.4.1: A_0 is covered if and only if the intersection points of all the arcs in A_0 are covered.

Proof: Suppose S_0 has overlapping regions of coverage with nodes $S_1, S_2, ..., S_n$. This means that there exist non-empty surfaces $sur_1....sur_n$ that partition A_0. Also, since all the arcs lie in A_0, the intersection of arc_{ij} and arc_{kl} will also lie inside A_0. Now, if S_0 is covered, then every point in A_0 is covered. Therefore, the intersection points of all the arcs in A_0 are covered.

To show the _"only if"_ part of the proof, suppose that all the intersection points are covered but there exists a point P in A_0 that is not covered. Since S_0 has overlapping coverage with adjacent nodes, every B_i will partition A_0 into a region covered by node S_i and a region external to A_i. Any point P in A_0 that is not covered must be external to every A_i and is hence bounded by a polyhedron whose sides belong to either sur_i $i=1..n$ or B_0. Since the sides of the polyhedron belong to the surfaces of the sensing regions, by definition 2.2 all the points on the surface of this polyhedron are also not covered. This implies that the vertices of the polyhedron, which also lie on the surface of the polyhedron bounding the point P, are also not covered. But, these vertices represent the intersection of arcs arc_{ij} and arc_{kl} for $i, j, k, l \in 1,..,n$ and are covered according to the assumption. This contradiction implies that if all the intersection points are covered, then there are no uncovered points in A_0, i.e., A_0 is completely covered by the sensor nodes $S_1, S_2, ..., S_n$. □

Algorithm for obtaining reduced deployment

Theorem 3.4.1 can be used to develop an algorithm for determining if the sensing region of any node is covered by one or more of its neighbors. If true, then the node can be deactivated without affecting the coverage of the network. This process can be implemented iteratively to deactivate all the nodes that are covered by one or more of their neighbors thereby resulting in a reduced network. In order to develop the algorithm, it is first necessary to determine the neighbors of each node. The following definitions aid in this process.

Definition 3.4.1: Sensor node S_i is a neighbor of a sensor node S_j if and only if $d(X_i, X_j) \leq 2R_c$, where R_c is the communication radius of sensor nodes S_i and S_j. $\qquad \square$

Definition 3.4.2: The neighbor set $N(i)$ of sensor node S_i is the set of all the neighbors of node S_i and is defined as $N(i) = \{S_j \in C \mid d(X_i, X_j) \leq 2R_c ; j = 1,.., n, j \neq i\}$. $|N(i)|$ is the cardinality of set $N(i)$ i.e. the number of neighbors of the node S_i. $\qquad \square$

Reduced Coverage Algorithm:

Step 1: For each sensor node S_i, form the set of neighbors $N(i)$. For each element 'S_k' in $N(i)$ compute the volume of overlap, V_{ik}, between node S_i and S_k. If the total overlap between S_i and its neighbors is less than the volume of S_i, then S_i is not completely covered by its neighbors and must remain active. That is,

$$V_{overlap} = \sum_{k=1;k\neq i}^{n} V_{ik} \leq \frac{4}{3}\pi R_s^3 \Rightarrow S_i \text{ remains active.}$$

Step 2: $V_{overlap} \geq \frac{4}{3}\pi R_s^3$ does not necessarily mean that S_i is covered. To check if S_i is covered, for every pair S_j, S_k in $N(i)$ do the following:

a) Find the arc 'arc_{jk}' obtained by the intersection of the coverage surfaces of S_i, S_j, i.e. $sur_j \cap sur_k$

b) Find all the intersection points of $arc_{jk}, arc_{lm}, j,k,l,m \in 1,..,N(i)$.

c) If all the intersection points obtained in (b) are covered then deactivate S_i.

The main advantage of this algorithm is that it is low in computational complexity and is executed in a distributed manner. The algorithm requires that each sensor node knows the locations of all its neighbors. The neighbor list can be easily compiled by the nodes based on the 'HELLO' messages exchanged at start up. When a network is deployed, all nodes are initially active. As the algorithm progresses, redundant nodes will switch to the inactive mode until no more nodes can be turned off without causing a hole in the coverage. To avoid neighbor nodes running the algorithm simultaneously and causing a 'blind spot', each node announces to its neighbors that it is currently running the coverage algorithm. If the node is redundant and is eligible for turning off without affecting the overall coverage, it will broadcast a 'GOODBYE' message to its neighboring nodes. Neighboring nodes receiving such a message will delete the sender's information from their neighbor lists.

The complexity of the algorithm proposed can be calculated as follows. Let 'N' be the maximum number of neighbors for any node in the network, i.e. $N = \max\limits_{i=1}^{n} |N_i|$. Then for any sensor node in the network, an arc is obtained by considering the intersection of the boundaries of any two neighbors of the node. Thus, a maximum of $^{N}C_2$ arcs must be calculated. Since intersection points require the consideration of pairs of arcs, in the worst case, $^{(^{N}C_2)}C_2$ intersection points must be checked for coverage. Therefore for a network of 'n' nodes, the algorithm requires a maximum of $n^{(^{N}C_2)}C_2$, i.e.

$$\frac{N(N-1)(N^2 - N - 2)}{8} n$$ steps in order to determine the reduced coverage. It is well known that the coverage problem is NP-hard [77]. For large networks, the number of neighbors of any node is small compared to the size the network $(N \ll n)$. The computational complexity of the algorithm for such large networks is of order 'n' ($O(n)$) where n is the total number of nodes in the network.

While the results presented in this section make possible the selection of a subset of sensor nodes in a WSN to cover a region, the result is a reduced cover but not necessarily an optimum cover for the region. Further, since the algorithm does not produce a unique result, it is advantageous to have a performance measure for comparing two different collections of sensor nodes that cover a region.

Definition 3.4.3: The **measure of optimality** of a WSN is the ratio of the number of active nodes in the network to the minimum number of nodes that can completely cover the same region. $\square$

The previous results show that the optimum deployment of sensor nodes in 3D would result in nodes located at the vertices of a _bcc_ lattice. Therefore, given the region to be monitored, one could easily find the minimum number of nodes required and their location for complete coverage. However, if the nodes are already deployed and a subset of these nodes selected to keep active, then the _measure of optimality_ is a measure of excess energy spent in monitoring the region as compared to an optimum deployment of the sensor nodes. A network with a lower '_measure of optimality_' would result in lesser expenditure of energy in monitoring the region.

3.5 Another Coverage Algorithm

In this section, another coverage algorithm that will serve as performance measure for the proposed coverage algorithm in Section 3.4 is proposed.

<u>Algorithm 2:</u>

Let $C_n = \{S_0, S_1, S_2, ..., S_n\}$ be the set of all the sensor nodes that cover the region **R**. Let the intersection of sphere S_0 and sphere S_i be given by the circle C_i, i.e. $C_i = A_0 \cap A_i$. (see figure 4). The interior of the circle C_i is said to be the disc bounded by the circle C_i, i.e. $D_i = interior(C_i)$. A circle C_i is completely covered if the disc bounded by the circle is completely covered, i.e. $\forall p \in D_i, p \in \overset{n}{\underset{j=1}{\cup}} A_j$.

<u>*Theorem 3.5.1*</u>: Let the sensor node S_0 be adjacent to sensor nodes $S_1, S_2, ..., S_n$ and $C_k = A_0 \cap A_k, k = 1..n$ be the circles of intersection. S_0 is completely covered if and only if all C_k's, $k = 1..n$ are covered.

<u>*Proof:*</u> If S_0 is completely covered, then every disc D_k in S_0 is covered. Definition of complete coverage of a disk ensures that the "*if*" part of the theorem holds.

To show the "*only if*" part, consider the following proof: If a circle is completely covered then each spherical segment adjacent to it is also covered. Each spherical segment must be bounded by some circle segment, and since each circle C_i is completely covered then all spherical segments are also completely covered. $\square$

The utility of Theorem 3.5.1 is in reducing the coverage of sensor node S_0 to a simpler problem of checking if all the circles of intersection between S_0 and the adjacent nodes are covered. This still is a complex task that is difficult to achieve in real time. The following theorem demonstrates a technique where such coverage can be determined by a few straightforward computations.

Theorem 3.5.2: A circle C_0 is completely covered by spheres if all the intersection points $C_i \cap C_j \in D_0, \forall i, j = 1...n$ are covered by one or more adjacent spheres.

Proof: Consider an uncovered point 'p' in D_0. Since some parts of D_0 are covered by adjacent sensor nodes, these spheres are going to partition D_0 into regions bounded by arcs from the boundary of C_0 and/or arcs from circles C_k's, $k = 1,...,n$. Suppose 'p' belongs to a region R_x in D_0. Since 'p' is not covered, is easy to see that R_x has to be bounded only by the exterior arcs of the circles. Also, the entire boundary of R_x, including the intersection points of the arcs, must have the same coverage status as 'p', i.e. all the intersection points on the boundary of R_x in D_0 must not be covered. This contradicts the assumption that all the intersection points $C_i \cap C_j \in D_0, \forall i, j = 1...n$ are covered. Therefore, if all the intersection points $C_i \cap C_j \in D_0$ are covered by one or more adjacent spheres, then D_0 is covered. Consequently, C_0 is covered. $\qquad\qquad \square$

Coverage Algorithm 2

Theorem 3.5.1 and 3.5.2 indicate that a sensor node is completely covered if all the intersection points $C_i \cap C_j \in D_k$ are covered by some sensor $S_l, l \neq i, j, k = 1..n$. Therefore, to check if S_0 is completely covered; one has to first find all the circles obtained by the intersection of $S_0 \cap S_k, k = 1..n$. For each C_k, find all the intersection points that lie within D_k. If all these intersection points are covered, then the circles C_k are covered. Then, by the theorems 3.5.1-3.5.2, S_0 is covered and can be deactivated. The following definitions aid in this process.

Steps in the distributed coverage algorithm:

Step 1: For each node S_i, form the set of neighbors

$N(i)$. For each element 'S_k' in $N(i)$ compute the volume of overlap, V_{ik}, between sensor node S_i and S_k. If the total overlap between S_i and its neighbors is less than the volume of S_i, then S_i is not completely covered by its neighbors and must remain active. That is

$$V_{overlap} = \sum_{k=1}^{n} V_{ik} \leq \frac{4}{3}\pi R_s^3 \Rightarrow S_i \text{ remains active.}$$

Step 2: $V_{overlap} \geq \frac{4}{3}\pi R_s^3$ does not necessarily mean that S_i is covered. To check if S_i is covered, for every pair of nodes S_j, S_k in $N(i)$ do the following:

    d) Find C_{ij} the circle got by the intersection of the coverage surface of S_i, S_j.

    e) Find C_{ik} the intersection circle of S_i, S_k.

f) Find the intersection points $C_{ij} \cap C_{ik}$.

g) If the intersection points are all covered, i.e. $C_{ij} \cap C_{ik} \in A_l, S_l \in N(i), l \neq i, j, k$, then deactivate S_i.

The computational complexity of the algorithm developed in this section is $O(N^3)$ where $N = \left(\overset{n}{\underset{i=1}{max}} |N(i)| \right)$ is the maximum number of nodes in the neighbor set of any sensor in the network.

3.6 Connectivity in 3D Wireless Sensor Networks

Consider a WSN comprising of sensor nodes $S_1,...,S_n$ with sensing radius R_s, and communication radius R_c, respectively. Let **R** be the region to be monitored. Denote the sensing and communication regions of a sensor node S_i as A_{is} and A_{ic} respectively.

Let $\overline{C} = \cup_{i=1}^{i=m} A_{is}$ and $\overline{\overline{C}} = \cup_{i=1}^{i=m} A_{ic}$ be the sensing space and communication space of a set of nodes $C = \{S_1, S_2,...S_m\}$ respectively.

Definition 3.6.1: A set of nodes $C_a = \{S_{a1},...,S_{am}\}$ is said to be connected if for every $a, b \in \overline{\overline{C}}_a$, there exists a continuous function $f:[0,1] \to \overline{\overline{C}}_a$ such that, $f(0) = a; f(1) = b$, and $\forall t \in (0,1)$, $f(t) \in \overline{\overline{C_i}}$. ☐

Connectivity implies that the location of any active sensor node is within the communication range of one or more active nodes such that all the active nodes form a connected communication backbone, while coverage requires all locations in the coverage region to be within the sensing range of at least one active node. Obviously, the relationship between coverage and connectivity depends on the ratio of sensing radius to communication radius. In the two dimensional case, it was shown that coverage implies connectivity whenever the radius of communication is twice the radius of coverage [18]. This result can be trivially extended for coverage and connectivity in three dimensions.

Lemma 3.6.1: A necessary and sufficient condition to ensure that coverage implies connectivity in 3D is that the radius of communication be at least twice the sensing radius. ☐

If the condition of Lemma 3.6.1 holds, then complete coverage automatically implies connectivity. Therefore, the connectivity problem is not explicitly addressed in this formulation.

3.7 Simulation Results

The theoretical developments in Sections 3.3-3.6 are validated through numerical examples in this section. First the case of random deployment of sensor nodes is studied and compared to the optimum deployment. Both 2D and 3D cases are considered and the number of nodes required for coverage studied. In the second case, given a deployment of sensor nodes, a reduced cover is obtained.

Example 1: In this example, the number of sensor nodes required to cover a 2D region of size 10 units by 10 units is studied. Random deployment, as well as deployment using square and hexagonal lattices is studied for different values of the sensing radius. Table 1 shows the results for values of sensing radius ranging from 2 to 0.3 units. It can be seen that the Hexagonal Lattice Deployment always results in the least number of sensor nodes. It can also be seen that the number of nodes required increases as the sensing radius decreases. In the case of random deployment, the number of nodes required for coverage increases exponentially as the sensing radius decreases. Figure 3.3 shows the number of nodes required for coverage on the sensing radius. Since low cost, low power sensor nodes typically have small sensing radius, random deployment of such nodes requires the use of a very large number of nodes. Therefore, efficient deployment of such sensor nodes requires a technique for determining a reduced subset of sensor nodes from a randomly deployed set of nodes.

Table 3.1: Number of sensor nodes required for coverage of a 10x10 region under different deployment strategies.

Sensing Radius	Random Deployment	Square Lattice Deployment	Hexagonal Lattice Deployment
2	37	16	12
1.5	72	25	20
1	150	64	42
0.6	367	144	120
0.3	1355	576	449

Table 3.2: Number of sensor nodes required for coverage of a 10x10x10 region under different deployment strategies.

Sensing Radius	Random Deployment	*bcc* lattice Deployment
2	182	54
1.5	325	128
1	773	396
0.6	2228	1421
0.3	24551	10952

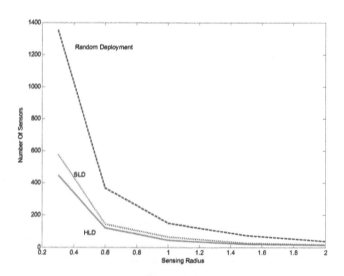

Figure 3.3: Number of sensor nodes (with different sensing radii) required for coverage of a 10x10 region under different deployment strategies: Random Deployment, Square Lattice Deployment (SLD) and Hexagonal Lattice Deployment (HLD)

Example 2: In this example, the number of sensor nodes required to cover a 3D region of size 10x10x10 units is considered. Table 3.2 shows the comparison of the number of nodes required for coverage using random deployment and deployment using a bcc lattice. In Figure 3.4, the required number of nodes with different sensing radii using random deployment and bcc lattice deployment are compared.

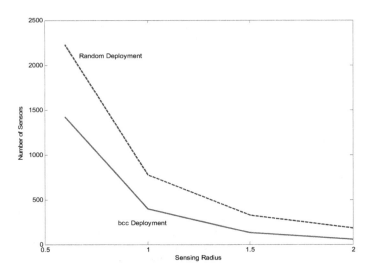

Figure 3.4: Number of sensor nodes required (with different sensing radii) for coverage of a 10x10x10 region under different deployment strategies: Random Deployment and *bcc* lattice deployment with different sensing radii.

In Examples 1 and 2, the sensor nodes were distributed randomly using a uniform distribution over the entire region of interest. While square lattice deployment has been widely used in the literature, it can be seen that hexagonal distribution in 2D and the *bcc* lattice in 3D, result in the optimum number of nodes for coverage. The structures obtained under hexagonal lattice deployment and *bcc* lattice deployment are shown in Figure 3.5(a) and Figure 3.5(b) respectively. The results in Table 3.1 show that in order to cover a planar region, fewer sensor nodes are required using hexagonal lattice deployment as compared to the popular square lattice deployment. Figure 3.5(b) gives some insight into the placement of sensor nodes in three dimensional deployment of WSN especially in buildings and in underwater applications [7], [8].

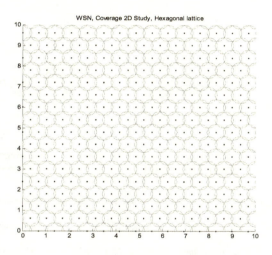

Figure 3.5(a): Deployment strategy in 2D using hexagonal lattice arrangement of 247 nodes. The coverage region is 10 units x 10 units and the sensing radius is 0.4 units.

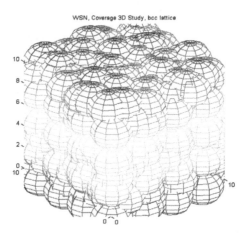

WSN, Coverage 3D Study, bcc lattice

Figure 3.5(b): Deployment strategy in 3D using *bcc* Lattice arrangement of 128 nodes. The coverage region is 10 units x 10 units x 10 units and the sensing radius is 1.5 units.

Example 3: In this example, the optimum coverage algorithm described in Section 3.4 is used to find the reduced cover of region 10x10x10 units when sensor nodes are randomly deployed. The nodes have a sensing radius of 1 unit and initially 2000 nodes are randomly deployed in this region using a uniform distribution. Figure 3.6(a) shows the initial deployment of the nodes and Figure 3.6(b) shows the reduced cover obtained by the algorithm in Section 5. It can be seen that 422 nodes were active in the reduced cover resulting in savings of 78.9%. If a *bcc* lattice deployment was used, then the minimum number of sensor nodes required would be 396. Therefore, the reduced cover has a *measure of optimality* equal to 1.06, which indicates that the algorithm has resulted in a solution very close to the optimum cover that is obtained by *bcc* lattice deployment.

89

To show that the simulation experiment wasn't simply a lucky situation, we compare the size of the reduced cover with different numbers of deployed sensor nodes which resulted in almost the same number of active sensors as shown in Figure 3.6(c).

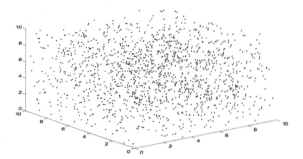

Figure 3.6(a): Random distribution of 2000 sensor nodes over a region 10x10x10 units.

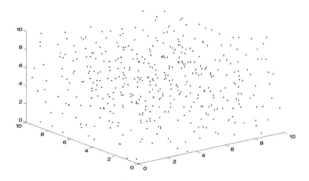

Figure 3.6(b): Reduced cover of the region 10x10x10 units. Initial Deployment = 2000; Number of active nodes = 422.

90

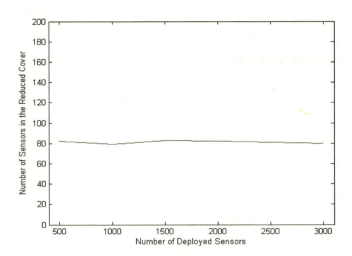

Figure 3.6(c): Different number of deployed sensor nodes resulting in different sizes of the reduced cover close to the optimal size from the optimal deployment strategy.

Example 4: The goal of this example is to compare the occupancy matrix before and after running the coverage algorithm. In this example, the optimum coverage algorithm described in Section 3.4 is used to find the reduced cover of region 10x10x10 units when sensor nodes are randomly deployed. The nodes have a sensing radius of 2 units and initially 2000 nodes are randomly deployed in this region using a uniform distribution. A 3D grid (grid size = one unit) is generated and 1000 grid points are tested for coverage. Figure 6 shows the occupancy matrix of each grid point in the initial deployment of the sensor nodes compared to that after running the coverage algorithm. Table 3.3 displays the minimum and maximum number of sensor nodes covering a cell in the grid before and after running the coverage algorithm. It can be seen that after running the algorithm a maximum of 5 sensor nodes cover any cell in the grid. The degree of cover after running the algorithm is 1, which guarantees that the region is completely covered.

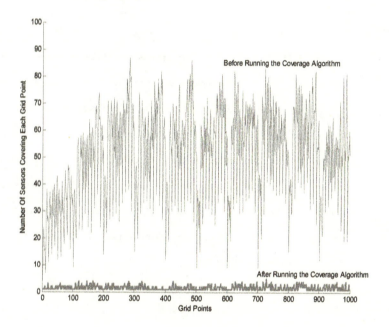

Figure 3.7: 1000 Grid points were tested, Sensor Radius=2, Grid Size =1, Occupancy Matrix - Initial Deployment (top) and after running the algorithm (bottom).

Table 3.3: Comparing the minimum and maximum cover before and after running the algorithm

	Min. Cover	Max. Cover
Before	8	89
After	1	5

Example 5: In this example, the minimum number of nodes that are required for random deployment is studied. In Example 3, 2000 nodes each with a sensing radius of 1 unit, were randomly deployed to cover a region of 10x10x10 units. The optimum cover of this region required 396 nodes resulting in 1,604 inactive nodes. In order to determine the number of sensor nodes required to cover the region, a number of trials were conducted and the resulting coverage was analyzed. The number of holes in the coverage for 10 trials is shown in Figure 3.8. It is seen that at least 800 nodes are required to adequately cover the region when random deployment is used. The reduced cover algorithm therefore results in a saving of 48%.

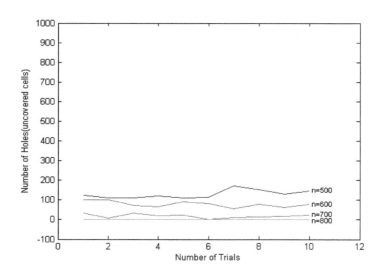

Figure 3.8: 10 different trials of a random 3D deployment of sensor nodes on a 10x10x10 region and the resulting holes in the coverage.

Example 6: The reduced cover of a deployment is obtained by first establishing communications between nodes to establish the list of neighbors and then iteratively executing the algorithm at each node. Therefore, the communication overhead increases as the number of deployed nodes increase. Figure 3.9 shows the average number of messages over 10 trials for different number of deployed nodes. From this figure, it can be seen that the communications overhead for the initialization of the network increases exponentially with the increase in deployed nodes. The advantage of using a reduced cover is further demonstrated by comparison of the performance of the flooding algorithm both in the original deployment, as well as in the reduced WSN. The number of messages for different queries is shown in Figure 3.10. It can be seen that the flooding algorithm results in a fewer number of messages in the reduced network.

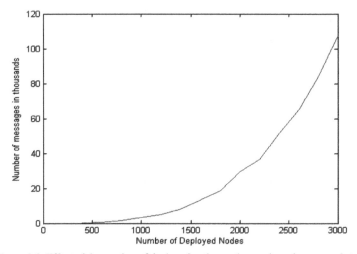

Figure 3.9: Effect of the number of deployed nodes on the number of messages between the nodes.

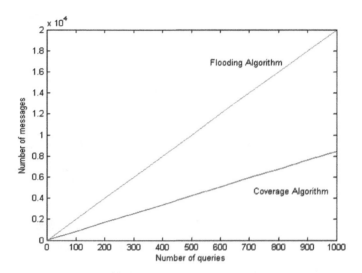

Figure 3.10: Number of queries vs. number of messages in a WSN. (i) Initial deployment (Flooding Algorithm); (ii) Reduced deployment (Coverage Algorithm).

The performance of the over time was also studied to determine the benefits of using a reduced cover. This is done by assuming that each sensor node has a limited energy supply of 300 Joules and is deactivated when the available energy is used up. The performance is evaluated in terms of *coverage lifetime*. The *coverage lifetime* is the continuous operational time of the system before the coverage drops below a specified threshold (for example 0.8). The nodes are assumed to consume 1400mW for each transmission and 1000mW for each reception. Further, the nodes are assumed to consume 830mW, 130mW in the idle and sleep states respectively. Figures 3.11(a) and 3.11(b)

96

show the overall coverage obtained over time as the WSN processes a series of queries. In these figures, two different initial deployments of 800 and 1600 nodes are considered. It can be seen in both the cases that the overall coverage drops over time as the available energy is used in processing the queries. Using the reduced network, it is seen that the resultant cover over time is significantly better. This is because each node in the reduced network has fewer neighbors and as a result has more efficient communications and less energy expenditure per query. This improvement in the coverage lifetime comes at a cost. The algorithm for obtaining the reduced network requires the communication between a node and its neighbors and as a result a portion of energy is used up during the initialization stage of the network. This causes early onset of degradation and loss of cover. This, however, can be addressed by incorporating *self healing* in the WSN.

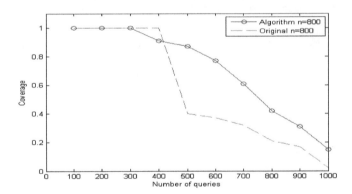

Figure 3.11(a): The effect of number of queries on the coverage lifetime of the WSN with 800 nodes.
(i) Initial deployment (Original); (ii) Reduced deployment (Algorithm).

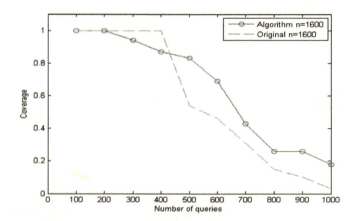

Figure 3.11(b): The effect of number of queries on the coverage lifetime of the WSN with 1600 nodes. (i) Initial deployment (Original); (ii) Reduced deployment (Algorithm).

To demonstrate the effect of *self healing*, a simple mechanism is implemented where a failing node in a reduced network alerts its neighbors about the impending failure. Inactive nodes in the neighborhood of the failing node are then activated. Since the inactive nodes and the failing node have overlapping cover, activating all the neighbors improves the coverage and results in better lifetime coverage for the WSN. The performance of the WSN for initial deployments of 800 and 1600 nodes with self healing is shown in Figures 3.12(a) and 3.12(b). Two different simulations are depicted in these figures, and in both the experiments the reduced network with self-healing can be seen to outperform the original network.

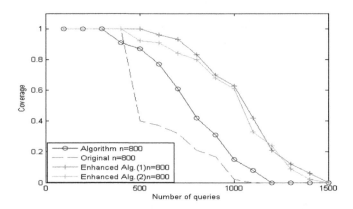

Figure 3.12(a): The effect of number of queries on the coverage lifetime of a self healing WSN with 800 nodes. (i) Initial deployment (Original); (ii) Reduced deployment (Algorithm).

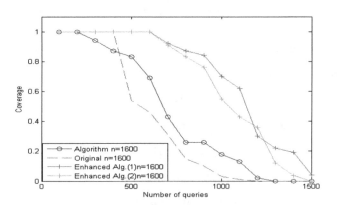

Figure 3.12(b): The effect of number of queries on the coverage lifetime of a self healing WSN with 1600 nodes. (i) Initial deployment (Original); (ii) Reduced deployment (Algorithm).

3.8 Chapter Conclusions

The main contribution of this chapter is a technique for obtaining a reduced cover of a wireless sensor network. The technique is shown to be computationally simple and suitable for distributed implementation. Numerical simulations show that the reduced sensor network has better energy efficiency compared to the random deployment of sensor nodes. It was demonstrated that the reduced WSN continues to offer better coverage of the region even when the sensor nodes start to fail over time. A localized 'self healing' algorithm is implemented that wakes up the inactive neighbors of a failing sensor node. Using the "flooding algorithm" for querying the network, it is shown that the reduced cover of the WSN with integrated self healing offers superior performance over time. For the first time, a 'measure of optimality' has been defined that enables the comparison of different implementations of a WSN from an energy efficiency stand point.

The proposed algorithm is simple computationally simple and will result in lower communication overhead. The 3D coverage algorithm can be easily extended to obtain application specific reduced cover, border coverage for intrusion detection, to determine the mobility of sensor nodes to cover sensing holes, and to incorporate self-healing in sensor networks.

In the next chapter, a more efficient self healing algorithm is developed and the notion of a 'substitute set' is introduced.

Chapter 4

Self Healing Coverage

In the previous chapter, the three dimensional coverage problem was studied and a distributed algorithm for complete coverage was provided. The simulation results show that the system lifetime could be extended if the coverage algorithms were used. However, sensor networks also introduce new challenges for fault-tolerance. Sensor networks are inherently fault-prone due to the shared wireless communication medium: message losses and corruptions (due to fading, collision, and hidden-node effect) are the norm rather than the exception. Moreover, node failures (due to crash and energy depletion) are commonplace. Since on-site maintenance is not feasible, sensor network applications should be self-healing. Another challenge for fault-tolerance is the energy-constraint in the sensor nodes. Applications that impose an excessive communication burden on nodes are not acceptable since they drain the battery power quickly. Thus, self-healing in sensor networks should be local and communication-efficient. This chapter proposes a solution for the design of a self-healing sensor network. We enumerate below the design goals for self-healing sensor network architecture.

1. *Complete Coverage*: The region should be completely covered at all time and the self healing algorithm should retain maximum coverage in the event of node failures.

2. *Optimality and Efficiency*: The algorithm should result in a reduced cover that keeps active the least number of sensors required for coverage. The reduced WSN should be as close to an optimal deployment as feasible in order to minimize the energy consumption of the nodes. The coverage algorithm must be independent of the topology of the WSN and implemented in a distributed manner. Further, the algorithm must be scalable for large sensor networks. The number of messages exchanged between adjacent sensor nodes must be kept to a minimum irrespective of the network size in order to guarantee efficient deployment of the WSN.

3. *Autonomy*: In remotely deployed sensor networks, a large degree of operational independence is essential. The network should be capable of self-organization to optimize energy usage by selecting a topology involving a minimum number of sensor nodes. Failure of an individual component, node or communication link should have minimal impact on the entire sensor network operation. The network must be able to detect local failures of a node and reorganize locally to guarantee maximal coverage of the region. This self healing should be accomplished with minimum computational complexity.

Self-healing at the hardware level is fairly common in both wired and wireless networks. Typically, if a piece of hardware absolutely cannot fail, a redundant, back-up system is installed and activated immediately in the event of an error. However, this redundancy is difficult to achieve in a low cost sensor node with limited on board energy.

Koushanfar et al. [78] solve this problem by adding redundant sensors. In the proposed system, they add an extra sensor to a system with N sensors in a way that any combination of N-1 sensors provides the desired result. Y. Zhang and K. Chakrabarty [79 and 80] describe a check pointing system for embedded systems in which sanity checks are performed at constant intervals. If a check fails, the node rolls the program back to the last successful checkpoint rather than having to rerun it from scratch.

In this chapter, we are more interested in self healing the network as a whole and not each sensor mote separately. The ultimate goal of this work is to improve the overall performance of the wireless sensor network comprised of sensor node. In the early 90s, Marzullo [81] was the first to address the problem of adapting to faulty sensor readings. The key idea is that if two sensors sample the same physical value, then their intervals must intersect. Marzullo's algorithms are centralized and not applicable to very large scale systems. The authors develop a multi-modal sensing approach to fault-tolerance. If one type of sensor fails in the environment, the application can dynamically activate the other sensor. Krishnamachari and Iyengar [82] have proposed a solution to the recognition of faulty sensor readings, and introduced algorithms for self-organization which combine shortest-path routing, and the construction of a spanning tree as a clustering mechanism for nodes in a feature region.

The rest of the chapter is organized as follows. A self healing sensor cover is introduced is Section 4.1 and three approaches to the self healing coverage problem are introduced. Simulation results that validate out theoretical contributions are provided in Section 4.2. The chapter is concluded is Section 4.3.

4.1 Self Healing Sensor Cover

The algorithms developed in Chapter 3 can easily be extended to incorporate self healing features necessary for the robust operation of the network. The self healing coverage problem can therefore be addressed in the following manner.

Problem (Self-Healing): Give a reduced cover with some holes due to some node failures and deaths; modify the cover by activating some sleeping nodes in order to cover the coverage holes in the region of interest.

Fault-tolerance is the ability of a system to deliver a desired level of functionality in the presence of faults. Fault-tolerance is crucial for many systems and is becoming vitally important for computing- and communication- based systems as they become intimately connected to the world around them, using sensors and actuators to monitor and shape their physical surroundings.

For dealing with arbitrary corruptions, we need self-healing systems: A self-healing system ensures eventual satisfaction of system specifications upon recovering from a fault. However, since faults can temporarily violate the program specifications in a self-healing system, extra care should be taken for containing the effects of faults: Faults in one part of the system may contaminate the entire system and hence may lead to a high-cost, system-wide correction.

It is assumed that the region of interest is initially covered and the holes in coverage overlap as a result of node failures or migrations. Then, self-healing involves the determination of the set of nodes in the vicinity of the coverage hole. Activating these set of nodes can result in the covering of the hole and there by retain the overall coverage of the region. There are three ways to approach the problem of self healing coverage:

1) In order to cover the holes in the region, the reduced coverage algorithm is applied on the entire region with the remaining set of nodes in order to get a new reduced set cover.

2) We have some holes in the cover, so the reduced coverage algorithm is applied on the holes, and we wake up some sleeping nodes to cover the uncovered holes in the region.

3) Initially when the reduced coverage algorithm is applied, each node selected in the reduced cover is assigned a substitute set SUB. The function of the substitute set is to cover the sensing region of the selected node in case of a failure. That is for every node S_x with sensing region A_x in the reduced cover, the substitute set is constituted as follows: $A_x \subseteq \overline{SUB}(S_x) = \bigcup_{i=1}^{m} A_i \mid S_1,...S_m \in SUB(S_x)$.

In case the faulty sensor belongs to a region that was originally a 1-cover, then the *SUB* set only guarantees maximum coverage of the affected region A_x.

The first approach is not cost efficient and will consume a great amount of energy. The second and third approaches are motivated by the same principle. However, the reduced cover obtained using approaches 2 and 3 are definitely not the optimal reduced cover but are reasonably close to it as will be show later in the simulation studies. The SUB algorithm (Approach 3) is as follows:

- Apply the reduced cover algorithm on the required region.

- If a node S_x is selected as part of the reduced cover, apply the reduced algorithm on the sensing region of S_x and the SUB reduced cover of that region is the SUB set of S_x.

- Each SUB set periodically sends a message to its designated sensor node to see if it is still alive or is about to die. If a reply is not received successfully, the SUB set is activated and complete coverage of the region is established.

An example of the self healing algorithm is presented in Figure 4.1. The idea of the algorithm is to select a substitute set for each sensor node in the reduced cover that will cover its sensing region in case of death or failure. To select a substitute set of a sensor node, the coverage algorithm introduced in Chapter 3 is applied on the sensing region of each sensor node in the cover i.e. the region of interest **R** is now replaced by the sensing region A_i of sensor node S_i. Let 'N' be the maximum number of neighbors for any sensor in the reduced cover network, i.e. $N = \overset{n}{\underset{i=1}{max}}|N_i|$. The computational complexity of the substitute selection algorithm is simply $O(mN)$ where 'm' is the size of the reduced cover.

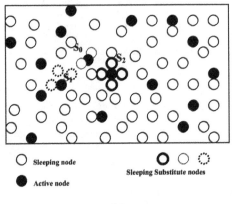

(a)

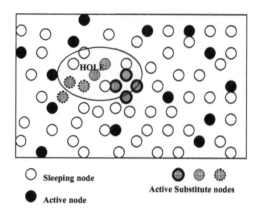

(b)

Figure 4.1: Illustration of the substitute plan algorithm for covering the holes (Approach 3). S_0, S_1, and S_2 were part of the sensor cover and after a period of time run out of energy. A hole in the region of interest is created. To cover the hole, S_0, S_1 and S_2's substitute nodes are activated.

4.2 Simulation Results

In this experiment, the optimum coverage algorithm described in Chapter 3 is used to find the reduced cover of region 10x10x10 units when sensor nodes are randomly deployed. The nodes have a sensing radius of 2 units and initially 1000 nodes are randomly deployed in this region using a uniform distribution. Some nodes are randomly disabled resulting in holes in the region (see Figure 4.2 (c)).The occupancy matrix of the region with holes is shown in Figure 4.3(a). The SUB algorithm described is Section 4.1 is used in order to obtain a new close to optimal reduced cover of the region (see Figure 4.2(d)). The occupancy matrix of the region after the new reduced cover is shown in

Figure 4.3(b). Comparing the SUB algorithm to the first approach which is simply recalculating the reduced cover of the entire region, it could be seen that the number of nodes in each reduced cover are very close and the energy wasting due to the computational complexity of the reduced algorithm is avoided by using the SUB algorithm (see Figure 4.4).

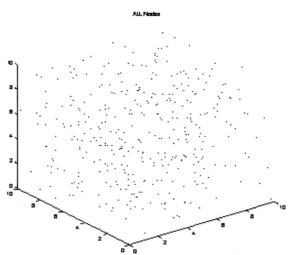

Figure 4.2(a): Random distribution of 1000 sensors over a region 10x10x10 units.

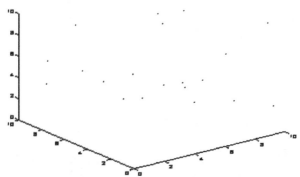

Figure 4.2(b): Reduced cover of the region 10x10x10 units. Initial Deployment = 1000; Number of active sensors = 79.

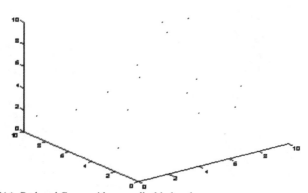

Figure 4.2(c): Reduced Cover with some disabled nodes.

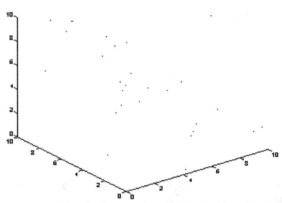

Figure 4.2(d): New Reduced Cover using SUB algorithm; Number of active sensors=111.

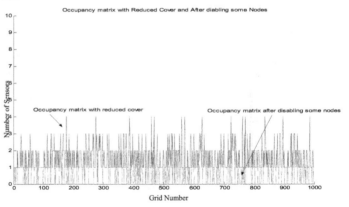

Figure 4.3(a): Occupancy matrix of the original reduced cover and after disabling some nodes.

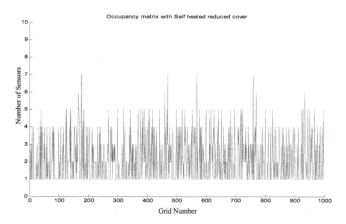

Figure 4.3(b): Occupancy matrix of the new reduced cover using SUB algorithm.

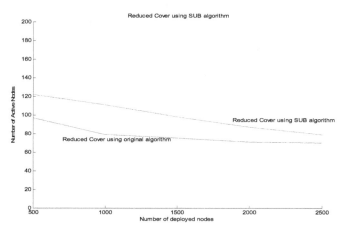

Figure 4.4: Reduced cover using SUB algorithm compared to the reduced cover using original algorithm with different number of deployed nodes.

4.3 Chapter Conclusions

In this chapter, unlike the previous work in this area, the coverage problem in 3-dimesnional wireless sensor networks (WSNs) was formulated and analyzed. A Self-healing algorithm was also established (which is an extension to the self healing algorithm introduced in Chapter 3). It was shown that this algorithm guarantees coverage at all time even if some nodes in the sensor cover run out of energy. The substitution plan results in great energy savings. For widespread adoption of the wireless sensor technology, robustness in the event of abnormal behavior such as a network intrusion, or failures of nodes is critical.

In the next chapter, the intrusion detection problem using wireless sensor networks is studied. Energy efficient approaches to the border coverage problem are proposed and simulations results are presented.

PART III

THE BORDER COVERAGE PROBLEM
IN WSNs

Chapter 5

Energy Efficient Approaches to the Border Coverage Problem in WSNs

In the previous 2 chapters (Chapter 3 and Chapter 4), the three dimensional full coverage problem in WSNs was studied and a self healing sensor cover was also provided. The focus of this chapter on the other hand, is the Boundary Coverage Problem which basically asks for the minimum subset of sensor nodes to guarantee the coverage of the boundary of a region of interest so that an intruder is detected at all times.

Networking together hundreds or thousands of cheap sensor nodes allows users to accurately monitor a remote environment by intelligently combining the data from the individual nodes. In this chapter, the *border coverage problem* in WSNs is rigorously analyzed. Most existing results related to the coverage problem in wireless sensor networks focused on planar networks; however, three-dimensional modeling of the sensor network would reflect more accurately real-life situations. Unlike previous works in this area, distributed algorithms that allow the selection and activation of an optimal border cover for both two dimensional and three dimensional regions of interest are provided. We also provide self healing algorithms as an optimization to the border coverage algorithms which allow the sensor network to adaptively reconfigure and repair itself in order to improve its own performance. Border coverage is crucial for optimizing sensor placement for intrusion detection and a number of other practical applications.

Numerical simulations show that the optimized border cover has better energy efficiency compared to the standard random deployment of sensor nodes interms of the overall system lifetime (Boundary Coverage). It is demonstrated that the optimized WSN with self healing enhancments continues to offer better border coverage of the region even when the sensor nodes start to fail over time.

The rest of the chapter is organized as follows. The motivation is introduced in Section 5.1. The boundary coverage and coverage boundary problems are formulated in Section 5.2. In Section 5.3, optimal 2D and 3D deployment strategies for border coverage are presented. In Section 5.4, distributed algorithms for identifying a minimal subset of sensor nodes that are on the exterior boundary of a given region are presented. In Section 5.5, algorithms for detecting sensor nodes that are on the exterior coverage boundaries as well as on the interior boundaries (coverage holes) are presented. In Section 5.6, a virtual borer patrol strategy is analyzed. Numerical simulation results that validate the proposed algorithms are presented in Section 5.7 and the chapter is concluded in Section 5.8.

5.1 Introduction and Motivation

Unsupervised intrusion detection, which involves detecting and identifying the encroachment of a monitored region by an object, is one of the applications of wireless

116

sensor networks. Algorithms for wireless sensor networks must have low communication overhead, rely as much as possible on local information, adapt to failures and changes in network conditions, and produce results in a timely fashion. Given the requirements to minimize the power, it is desirable to select the bare essential number of sensor nodes dedicated for the task while all other sensor nodes should preferably be in the hibernation or off state. Even though target tracking has been widely studied for sensor networks with large nodes and distributed tracking algorithms are available [50-60], intrusion detection in ad hoc networks with micro sensor nodes poses different challenges due to communication, processing and energy constraints.

Border surveillance is one of the major applications of sensor networks. The border represents the physical extent of the region to be monitored and depending on the application, it is required to sense the intrusion into the monitored region or exit from the monitored region of the object being monitored. In a typical deployment of sensor nodes, sensor nodes are distributed across the entire region of interest and it is necessary to determine a minimal set of sensor nodes that can adequately monitor the border. Thus, it is necessary to find a scalable and energy efficient solution to the border coverage problem. Such a solution would extend the scalability of wireless sensor networks and enable the monitoring of one of the largest international borders [83].

In this chapter, the problem of determining the minimum number of sensor nodes for covering the boundaries of a target region is addressed. Unlike the full coverage problem, here the primary interest is in the detection of movement of an object across the boundary. The problem of selecting the minimum subset of sensor nodes for covering the boundaries of a known target region with no coverage holes is first addressed (*Boundary Coverage Problem (BCP)*). However, the deployed sensor node density might not be dense enough to cover the boundary and the emergence of coverage holes in the target area is unavoidable due to the following reasons:

1- *Random Deployment*: Random deployment of sensor nodes is always desired; however it doesn't guarantee full coverage of the region of interest.

2- *Sensor Failures:* Nodes are subject to failures due to depleted batteries or, more generally, due to environmental influences. Sensors may fail from the impact of deployment or simply from extended use.

3- *Position Changing*: A lot of environmental factors (wind or storms) may change the sensor nodes' positions over time and possibly resulting in some coverage holes in the network. In addition to that, sensor nodes equipped with mobile capabilities might also result in some holes due to the sensor nodes frequent change in position.

4- *Presence of Obstruction:* Some obstacles in the region of interest might impair the nodes sensing/communication functionality and thus result in some coverage holes.

The second problem addressed in this chapter is the *Coverage Boundary Problem (CBP)* which is identifying the boundary of the wireless network's coverage region. At first glance, boundary detection is similar to edge detection in image processing. However, a major difference is that due to energy constraints, processing the entire image of the network at a single point is impractical and infeasible and hence a single node doesn't have all other sensor nodes' information. In this case, we differentiate between two kinds of coverage boundaries: exterior boundary (outer periphery of the network) and interior boundary (sensor nodes that define the coverage holes in the network). A wireless sensor network for detecting large scale phenomena may be called upon to provide a description of the boundary of the phenomena. Several phenomena (containment flows) can span large geographic areas. Sensing and detecting the boundary of the phenomena can help scientists understand what factors affect the spread of theses phenomena. A representation of the boundary of the coverage has the potential to be more concise and therefore more energy efficient that an enumeration of all the sensor nodes in the network for a specific query. We also argue that identifying coverage holes in the network, is not only used to detect regions with low sensor density due to depletion of node power (places where adding new nodes will significantly improve the coverage and connectivity of the network), but could also be used to identify the regions of interest for the end user. Identifying sensors nodes that are on the coverage boundary is motivated by a number of functionalities at both the network and application layers:

1- *Tracking:* A sensor network is used to track a moving intruder within a region of interest. Information about coverage holes can be used to help path establishment, where a communication path is to be kept between the moving objects and the sink.

2- *Routing:* Identifying the boundary of the both exterior and interior coverage is very helpful for routing algorithms where the data need to be sent to a specific region instead of a specific sensor node.

3- *Disaster Boundary:* In the case of a natural disaster (flood, hurricane, or fire) may lead to the destruction of all the sensors in the effected region and leave a coverage hole in the network. Detection of the boundary for this hole indicates the region of interest and the necessary response from the user.

4- *Monitoring Network Lifetime:* The sensor network is associated with its lifetime and therefore initiate any other necessary network management activities.

5- *Topographical Properties:* The detecting of the coverage holes is also a critical militarily application because it determines the ability of armed forces to take and hold areas, and to move troops and material into and through areas. Topographical properties can also be useful in determining weather patterns.

6- *Network Traffic:* Each sensor can be designed to keep a critical value which can be defined as the number of packets it has to send, or how busy its local wireless medium is. Real time traffic conditions can easily be used for routing or any kind of other applications.

The goal in this chapter is quite different from the ones mentioned above. We are more concerned with detecting the sensor nodes that are on the boundary of the region of interest (no coverage holes) or on the boundary of the coverage. Assuming the boundary of the region of interest is known; a distributed algorithm that selects a minimum subset of active nodes that will guarantee boundary coverage of the region is then developed. Since the emergence of coverage holes is unavoidable, distributed as well as centralized algorithms are provided to detect coverage holes in the region and also find the nodes that are on the boundary of the coverage area. The algorithms presented are distributed and are very low in computational complexity. A recent work related to the work presented in this book considered the hole coverage problems in a sensor network comprising of stationary nodes with minimal geometric data [84]. Here, the authors do not consider the location of the sensor nodes but use the concept of homology for detecting holes in the cover The computational complexity of the proposed algorithms and the execution of the algorithm in a distributed manner were not analyzed. In [85], Carbunar, *et. al.* study the problem of detecting and eliminating redundancy in a sensor network with a view to improving energy efficiency, while preserving the network's coverage. The detection of the coverage boundary was also attempted by reducing it to the computation of Voronoi diagrams. To our knowledge, no work has considered distributed coverage boundary detection in sensor networks. Nowak and Mitra [86] describe a scheme for estimating the boundary of a large scale phenomenon by aggregating readings along a predefined hierarchical structure within the network. The work in this book is complimentary to the results in [86] and the distributed algorithms can be used in a number of different boundary estimation applications (including those considered in [86]).

5.2 Border Coverage Problem Formulation

An emerging application area for sensor networks is intelligent surveillance and intrusion detection. Sensor nodes are randomly distributed in an area to be monitored. The ultimate goal is to detect an intruder target and alert the sensor nodes which are close to the predicted path of the target. However, minimizing the power consumed should be the most important design goal. The lifetime of the sensor network can be significantly extended by optimizing the energy consumption of each sensor node. The emergence of coverage holes in the target area is unavoidable and depending on the application, the coverage hole problem could be used in different ways.

We assume that any two nodes s_i and s_j can directly communicate with each other if their Euclidean distance is less that the communication range R_c. Although a network can be rendered useless if it looses its connectivity, we characterize the system lifetime by just observing the resulting boundary coverage. Zhang and Hou [24] showed that if the communication range is at least twice the sensing range, then complete coverage of a convex area implies connectivity among the nodes. Assuming the communication range is twice the sensing range ($R_c \geq 2R_s$), the theorems in [24] could be easily extended to handle boundary coverage of the region as well.

Definition 5.2.1: An **intruder** is any object that is subject to detection by the sensor network as it crosses the border. □

A reasonable assumption is made that no intruder is aware of the location of the deployed sensor nodes.

The following will give precise definitions to what a border of a region is.

Definition 5.2.2: Let **R** be a subset of the (2D or 3D) space. The point 'p' is said to be **near R** if every neighborhood of 'p' contains a point from **R**. i.e.

$\forall \varepsilon > 0, \exists x \in Ball(p, \varepsilon) \ and \ x \in R.$ □

In the definition above, $Ball(p, \varepsilon)$ means the set of all points whose Euclidian distance from p is less than ε.

Definition 5.2.3: The set of all points in **R** and near **R** is called the **closure** of **R** and is denoted by $cl(R)$ i.e. $cl(R) = (R) \cup \{All \ points \ near \ R\}$. □

Definition 5.2.4: The **border** of a region **R** denoted by $B(R)$ is defined as the set of all points that are common to **R** and its complement i.e. $B(R) = cl(R) \cap cl(\overline{R})$ where $\overline{R}$ is the complement of the region **R** i.e. all the points that don't belong to **R**.   □

According to definitions 5.2.1-5.2.4, a region is said to be boundary covered if and only if an intruder is always detected as it crosses the boundary of the region. A sensor is called a boundary sensor if its sensing region intersects the boundary of the region of interest.

Definition 5.2.5: A set of sensor nodes C_{Border} is said to be a **boundary cover** of a region **R** if every point on the boundary of **R** belongs to the sensing region of at least one sensor in C_{Border} i.e. $\forall p \in B(R)$, $p \in S_i$ for some $S_i \in C_{Border}$.   □

Definition 5.2.6: A set of sensor nodes $C_{Border,Reduced}$ is said to be a **reduced boundary cover** of a region **R** if $\forall p \in B(R)$, $p \in S_i$ for some $S_i \in C_{Border,Reduced}$ and no proper subset of $C_{Border,Reduced}$ is a boundary cover of **R**. i.e.

$C_{Border,Reduced} - S_l$, for any $S_l \in C_{Border,Reduced}$, is not a boundary cover of **R** .   □

Definition 5.2.7: A sensor node is called a **redundant sensor node** if its sensing region is completely covered by its neighboring sensor nodes. Deactivating a redundant sensor won't affect the overall full coverage of the region of interest. (Figure 5.1(a))   □

124

Definition 5.2.8: A sensor node is called a **redundant boundary sensor** node if the portion of the boundary covered by it is completely covered by its neighboring sensor nodes. (Figure 5.1(b))   □

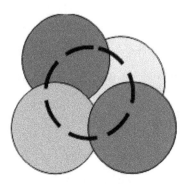

Figure 5.1(a): Example of a redundant sensor node (dashed circle).

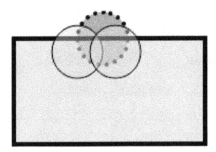

Figure 5.1(b): Example of a border redundant sensor (dashed circle) and a non-border sensor (black shaded circle).

Definition 5.2.9: A sensor node is called a **non-boundary sensor** node if its sensing region does not intersect the boundary of the region of interest. □

In the first case, it is assumed that each sensor node is aware of its own location, the location of the boundaries of the region to be monitored and the location of its neighbors. This assumption is not too stringent and it can be satisfied by communications between adjacent nodes in the network on startup. From definitions 5.2.8 and 5.2.9, it can be easily seen that the deactivation of a boundary redundant sensor node or a non-boundary sensor node will not affect the overall boundary coverage of the region of interest. Using definitions 5.2.1 – 5.2.9, the boundary coverage problem is analyzed in this chapter in the following way:

I. **Optimal Deployment for Border Coverage**: Find the minimum number of sensor nodes and their placements for border coverage of a given region **R**.

II. **Optimal Selection for Boundary Coverage**: Given a dense deployment of sensor nodes in a region **R** with known boundary, find a minimum subset of active nodes that guarantee boundary coverage of **R**.

In the second case, we are concerned with identifying the nodes on the boundary of the coverage as well as coverage holes in the region as depicted in Figure 5.7(b). A sensor S_i is not on the boundary of coverage (or coverage hole H) if and only if its sensing boundary circle Cir_i is completely covered by its neighboring sensors i.e. $S_i \notin B(H) \Leftrightarrow \forall p \in Cir_i, p \in A_j, \text{for some } S_j \in S$. Now, the boundary of coverage (or coverage hole H) can be defined in terms of a set of sensor nodes as:

Definition 5.2.10: The **coverage boundary** is defined as the set of all sensor nodes S_i such that: $\exists p \in Cir_i \mid p \notin A_j, \forall S_j \in S$.   □

The coverage boundary problem can now be addressed in the following way:

III. **Identifying Coverage Boundary:** Given a dense deployment of sensor nodes, find the subset of active nodes that lie on the boundary of the coverage area (or holes).

The discussion in the following sections assumes that the region to be monitored is large in comparison to the sensing region of an individual sensor node and that the location of all sensor nodes is known. All through the chapter, the following notations will be used:

R – Region of interest , S – Set of sensors in the region , R_s – Sensing radius of each sensor ,

A_i – Sensing region of sensor S_i , Cir_i – Boundary of the 2D sensing region of sensor S_i ,

C_{full} – Set of sensors fully covering the region , and C_{border} – Set of sensors border covering the region

5.3 Optimal Deployment for Border Coverage

When flexibility in deployment exists, it is advantageous to find an optimum border deployment of the sensor nodes so that border coverage can be achieved using a minimum number of nodes. In this section, theorems for optimal deployment of the sensor nodes are developed. These theorems provide lower bounds on the number of nodes needed to border cover both 2-dimesnional and 3-dimensional regions of interest.

5.3.1 Optimal 2D Deployment for Border Coverage

In the 2D deployment problem, the minimum number of sensor nodes modeled as disks and their locations for border coverage of a given rectangular region **R** are to be determined. While the region to be border covered is assumed to be a rectangular region, the algorithms could be easily extended to border cover any arbitrary shape of a region with minor modifications.

Lemma 5.3.1: Consider a rectangular region **R** of length 'L' and width 'W'. The lower bound on the number of sensor nodes needed to achieve border coverage of **R** is $2(\left\lceil \dfrac{L}{2R_s} \right\rceil + \left\lceil \dfrac{W}{2R_s} \right\rceil) - 3$ where $\lceil \ \rceil$ represents the operation of finding the least upper bound integer.

Proof: The optimal way to deploy the sensor nodes to achieve border coverage of the region is to deploy the sensor nodes across the perimeter of the entire region such that any 2 adjacent sensor nodes that are on the same row or column are tangent to each other. $\left\lceil \dfrac{L}{2R_s} \right\rceil$ is the least number of sensor nodes to cover a line of length 'L'. For a rectangular region of length 'L' and width 'W', the perimeter can be optimally covered by $2(\left\lceil \dfrac{L}{2R_s} \right\rceil + \left\lceil \dfrac{W}{2R_s} \right\rceil)$ sensor nodes. However such a cover will have overlapping sensing coverage at the vertices of the rectangle as shown in Figure 2. The number of sensor nodes doesn't exactly cover each edge then the last sensor would partly cover the adjacent edge, so a better way would be to select the next position of the center such that its circle intersects the last circle in its boundary intersection. Since the sensor nodes have equal sensing radii and their centers lie on the boundary of the rectangle then in the best case scenario, $\dfrac{R_s}{2}$ of the boundary line will be covered. The best enhancement on the first placement of sensor nodes would be minimizing the number of sensors by one on three boundary edges (Figure 5.2). So the lower bound on the number of sensor nodes to cover a 2D rectangular region each with sensing radius R_s is $2(\left\lceil \dfrac{L}{2R_s} \right\rceil + \left\lceil \dfrac{W}{2R_s} \right\rceil) - 3$. $\square$

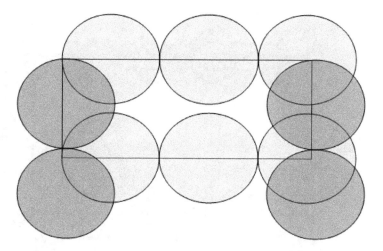

Figure 5.2(a): The optimal deployment of 10 sensor nodes modeled as circles in 2D to border cover a rectangular region.

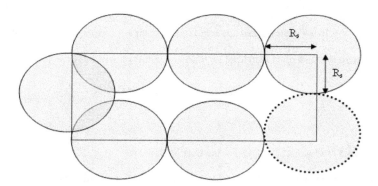

Figure 5.2(b): Illustration of the best possible way to minimize the number of sensor nodes covering the border. The deployment strategy is further enhanced to take advantage of a row sensor covering part of the column resulting in 7 deployed sensor nodes.

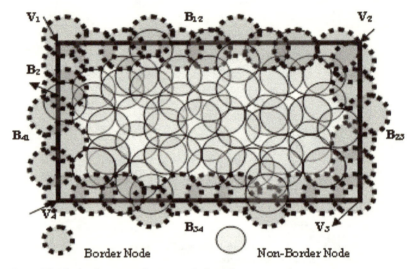

Border Node   Non-Border Node

Figure 5.3: The border cover of a rectangular boundary.

If we assumed that the region to be monitored is large in comparison to the sensing region of an individual sensor node, then the necessary and sufficient number of nodes to cover a rectangular region would simply be $2(\left\lceil \dfrac{L}{2R_s} \right\rceil + \left\lceil \dfrac{W}{2R_s} \right\rceil)$.

5.3.2 Optimal 3D Deployment for Border Coverage

The three dimensional optimal sensor deployment for border coverage is far more complex that the two dimensional case. It is addressed by determining the minimum number of sensor nodes required to cover the surface of a cubical region of interest. Since the coverage region of a sensor node is modeled as a closed ball, the border coverage problem requires the determination of all the points on the surface of the cube that are covered by the sensor nodes. To address this issue, the intersection of the sensing regions and a boundary plane is first defined. This definition will then be used to determine the least number of sensor nodes required for border coverage.

Definition 5.3.1: A **great circle** on a sphere is the intersection of that sphere with a plane passing through the center of the sphere. □

Lemma 5.3.2: The centers of all the optimal deployed spheres must lie on a face of the cube. □

Proof: It is clear that each sensor covers a maximum area when the coverage region lies in the plane passing through the center of the sphere representing the sensing region. Thus, minimizing the number of sensor required to cover the surfaces of the region to be monitored is equivalent to maximizing the coverage area of each sensor. This is possible only when the centers of all the sensor nodes lie on the surface of the region to be monitored. □

Lemma 5.3.1: The optimal deployment locations of sensor nodes to border cover a 3D cubical region is the locations of the spheres whose centers form a lattice of spacing $\Lambda = 1.7322 R_s$ on each face of the cube. □

Theorem 5.3.2: Consider a cubical region **R** of side 'a' ('a' is sufficiently large in comparison to R_s). An approximation on the lower bound on the number of sensor nodes of sensing radius R_s to achieve border coverage of **R** is $N_{min} = \left\lceil \dfrac{2.309a^2 - 3.845a}{R_s} \right\rceil$.

Proof: The proof is based on the problem of covering a square by circles which has been studied by Kershner [28] and Verblunsky [29] where N_c, the least number of circles of unit radius which can cover a square was determined. They proved that there is a constant $c \geq \dfrac{1}{2}$ such that for all sufficiently large 'a', $a^2 + ca < \dfrac{3\sqrt{3}}{2} N_c < a^2 + 8a + 16$. According to Theorems 3.1-3.3, the 3D border coverage problem was simplified to the problem of completely covering the faces of a cube with circles. Since there are 6 faces of a cube to be covered and we are interested in finding the lower bound of the number of sensor nodes needed, then $c = \dfrac{1}{2}$. Let R_s be the sensing radius of each sensor therefore a basic lower bound is $N_{min} = 2(\dfrac{2a^2 + a}{R_s \sqrt{3}})$. But since spheres can cover 2 or even 3 faces of the cube at the same time, the optimal way would be if the 2 faces intersect the sphere in semi great circles resulting in minimizing number of nodes by approximately $\dfrac{5a}{R_s}$ .Since we are concerned with a lower bound, $N_{min} = \left\lceil \dfrac{2.309a^2 - 3.845a}{R_s} \right\rceil$ is a valid lower bound on the three dimensional border cover.   □

5.4 Optimal Selection for Border Coverage

The results in section 5.3 enable the optimum placement of sensor nodes for border coverage of a given region (2D and 3D regions). In practice, however, given an existing distribution of sensor nodes, it is often necessary to minimize the number of nodes that remain active while still achieving border coverage of the entire region. In this section, an algorithm is developed where the nodes make local decisions on whether to sleep or join the set of active nodes. The two-dimensional and three-dimensional cases for selecting an optimum border cover of a given region are studied. A measure of optimality is also proposed to compare the performance of the border coverage of a given sensor network with the optimum coverage obtained in Section 5.3. The border coverage algorithm presented in this section has the following key features:

1. It is a decentralized algorithm that depends only on the local states of the sensing neighbors.
2. It provides guaranteed degrees of border coverage.
3. It handles the case where the nodes have unequal sensing radii.

5.4.1 A 2D Distributed Border Cover Selection Algorithm

In order to solve the border coverage problem for a two-dimensional region of interest, it is assumed that the region to be monitored is a rectangle specified by its vertices

$V_1, V_2, V_3, and V_4$. It is also assumed that all the sensor nodes are aware of the location of the vertices, i.e. the sensor nodes are aware of the extent of coverage that is required. The border coverage algorithm can be applied to any shape of boundary but the region of interest is assumed to be a rectangular region for the sake of ease of presentation. The algorithm depends on the fact that individual sensor nodes can verify if they have overlapping border coverage with their neighbors. If the border covered by a sensor node is covered by other sensor nodes in the neighborhood, then deactivating this sensor will not affect the overall border coverage. In this section, we will first derive conditions that indicate overlapping border cover for a given sensor. We will start by giving some definitions and assumptions that will aid us in developing an algorithm to select a border cover.

Let $B(\mathbf{R})$ represent the boundary of the region $\mathbf{R}$ to be covered. Then $B(\mathbf{R})$ can be represented $B(\mathbf{R}) = \bigcup_{\substack{i,j=1 \\ i \neq j}}^{4} B_{i,j}$ where $B_{i,j}$ is the segment connecting vertices V_i and V_j.

Without loss of generality, suppose the boundary edges are ordered as $B_{12}, B_{23}, B_{34}, and B_{41}$ as shown in Figure 5.4(b).

Definition 5.4.1: An **intersection segment** is the portion of the boundary covered by the sensing region of a sensor node and is represented by the closed interval [x, y] such that:

$[x, y]$ _is an intersection segment_ $\Rightarrow \forall z \in [x, y], \exists i \in 1,..,n$ such that $z \in A_i$ and $x, y \in Cir_i$

135

A segment $Seg_i = [x_i, y_i]$ is represented by its start point x_i and end point y_i. Examples of intersection segments are shown in Figure 5.4(a).

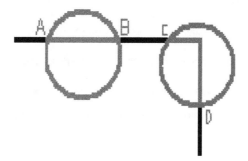

Figure 5.4(a): Intersection segments [A, B] and [C, D].

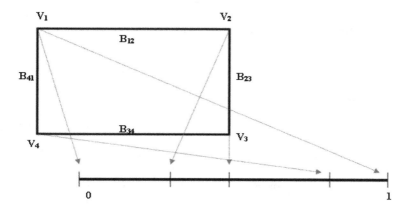

Figure 5.4(b): The ordering based on the mapping function.

Since the algorithms depend on the concept of ordering, a mapping: $\varphi: B \rightarrow [0,1]$ is defined based on the distance metric from the nearest origin i.e. $\forall x \in B_{ij}, \varphi(x) = \dfrac{d(x,V_i)+d(V_1,V_i)}{|P|}$ where $|P|$ is the total length of the perimeter of the rectangular boundary and $d(V_i,V_j)$ is the distance along the boundary of the region i.e. for example (see Figure 5.5) $d(V_1,V_3) = |B_{12}| + |B_{23}|$. A special case should be taken for the sensing region covering the origin vertex v_1, where the resulting intersection segment is divided into 2 sub-segments each of which is mapped separately.

Figure 5.5: Example of a segment and its successor, $[A,B] \succ [C,D]$.

Definition 5.4.2: We call $Seg_j = [x_j, y_j]$ the **successor** of $Seg_i = [x_i, y_i]$ denoted by $Seg_i \succ Seg_j$ if the following conditions are satisfied:

➤ $Seg_j \cap Seg_i = Seg_{ji} \neq \varnothing$

➤ $x_j > x_i$ & $y_j > y_i$

➤ there is no other starting point in Seg_{ji}. i.e. $\forall p \in Seg_{ji}, p \neq x_k$ for some $k \neq i,j$ □

Theorem   5.4.1:   Consider   the   set   of   segments

$S = \{Seg_1, Seg_2, ..., Seg_m\}$ _where_ $Seg_i = [x_i, y_i] \subset [0,1]$. Assume that no two segments are

contained in each other i.e. $\forall Seg_i, Seg_j \in S, Seg_i \not\subset Seg_j, i \neq j$. A segment $seg\ (= [x, y])$ is

covered by $\overset{m}{\underset{i=1}{\cup}} Seg_i$ if and only if the following hold:

    a.   there exist integers $1 \leq a, b, ..., k \leq m$ such that $x \in Seg_a$ _and_ $y \in Seg_b$,

    b. $Seg_a \succ Seg_b \succ ..... \succ Seg_k$.

Proof: $Seg_a \succ Seg_b \Rightarrow Seg_a \cup Seg_b = [x_a, y_b]$. Therefore, $\overset{k}{\underset{i=a}{\cup}} Seg_i = [x_a, y_b]$. Further from (a),

$x_a < x$, and $y_k > y$. $[x, y] \subset [x_a, y_k] = \overset{k}{\underset{i=a}{\cup}} Seg_i \subset \overset{m}{\underset{i=1}{\cup}} Seg_i$. On the other hand, suppose that the

segment [x, y] is covered by the segments $Seg_1, ..., Seg_m$. Since the segment is covered,

there exists some segment Seg_a such that $x \in Seg_a$. Similarly, there exists at least one

segment 'k' such that $y \in [x_k, y_k]$. Thus, condition (a) is easily satisfied. Now, if $y_a > y$,

then $[x, y] \subset Seg_a$ and condition (b) is trivially satisfied. Otherwise, there exists a

segment Seg_b such that $Seg_a \succ Seg_b$. If this was false, then it means that

$y_a \geq y_k$ _for_ $1 \leq k \leq m$. This would then imply that there exist points in the interval $(y_a, y]$ that

are not covered, thereby contradicting the assumption that the segment [x, y] is covered.

If $y_b > y$, then condition (b) is proved. Otherwise, repeating the process, we obtain

integers $a...k$ such that $Seg_a \succ Seg_b \succ ..... \succ Seg_k$ and $y_k > y$.

Therefore, conditions (a) and (b) together imply that the segment $[x, y]$ is covered by the collection of segments $Seg_1, Seg_2, ..., Seg_m$. $\square$

Theorem 5.4.1 indicates that a sensor node is completely border redundant if each segment in the partitioning of its intersection segment by its neighbors' intersection segments has a successor and the end points are also covered. Therefore, to check if a sensor S_0 is a border redundant sensor and therefore could be deactivated without affecting the overall border coverage, one has to first find all the adjacent sensor nodes that lie on the border of the region of interest. For each sensor, find the resulting intersection segment (or segments) with the boundary lines and check if S_0's portion of border coverage is completely covered by its neighboring sensor nodes. That can be done by using Theorem 5.4.1. An algorithm is presented that illustrates the steps in this process.

2D Distributed Border Coverage Algorithm

For each node S_i, form the set of neighbors, $N(i)$. Do the following:

Step 1: *Find the intersection segments* Seg_i

Find 'Seg_i' the intersection segment of S_i with the boundary of the region of interest and map it to [0, 1].

If $Seg_i = \varnothing$ or a point then S_i is declared as a non border node.

Else go to step 2.

<u>*Step 2*</u>: *Non containment property*

Let $\overline{Seg_i}$ be the set of segments covering Seg_i and is initially set to $\varnothing$.

For every pair of nodes s_j, s_k in $N(i)$

- Find the common intersection segments $Seg_j = [x_j, y_j]$ and $Seg_k = [x_k, y_k]$ resp.

- If the end points appear in increasing order as x_j, x_k, y_k, y_j i.e. $Seg_k \subseteq Seg_j$ and can be ignored.

- Update $\overline{Seg_i}$ to include Seg_j i.e. $\overline{Seg_i} = \overline{Seg_i} \cup \{Seg_j\}$

<u>*Step 3*</u>: *Check for endpoints coverage*

Check that, $\exists\, Seg_f = (x_f, y_f)$ *and* $Seg_l = (x_l, y_l)$ in $\overline{Seg_i} \mid x_f \le x_i \le y_f$ *and* $x_l \le y_i \le y_l$.

If true go to step 4.

<u>*Step 4*</u>: *Check for successor*

Check that, for each element $Seg_m = (x_m, y_m)$ in $\overline{Seg_i} - Seg_i$, $\exists\, Seg_n = (x_n, y_n) \mid$ $Seg_m \succ Seg_n$ and $m \neq n$.

If this condition is satisfied, the boundary intersecting segment Seg_i of the given sensor is completely covered and S_i is declared as a border redundant sensor and can be deactivated without affecting the overall border coverage. The algorithm guarantees that every point on the boundary of the target region is covered by at least one sensor. The optimal set of sensor nodes is also selected. The computational complexity of the redundancy selection algorithm developed in this section depends on $N = \left(\overset{n}{\underset{i=1}{max}} |N(i)| \right)$,the maximum number of nodes in the neighbor set of any sensor in the network and n , the total number of sensor nodes in the network. The computational complexity of the border redundancy checking algorithm is $O(N^2)$. Since 'n' sensor nodes need to be checked, then the complexity is $O(n.N^2)$. For large networks, the number of neighbors of any sensor is small compared to the size the network $(N \ll n)$ so the computational complexity of the algorithm for such large networks is of order 'n' ($O(n)$) where n is the total number of sensor nodes in the network.

5.4.2 A 3D Distributed Border Cover Selection Algorithm

The three dimensional optimal sensor border coverage problem is far more complex than the two dimensional case. We will approach it from a different angle and try to transform it to optimal complete coverage of the sensor nodes in a 2D plane. We will start by proving some theorems and then provide a 3D distributed algorithm.

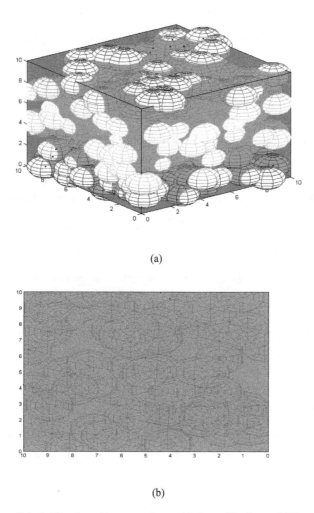

(a)

(b)

Figure 5.6: A 3D cube with some spheres. (a) One of its faces. (b)The corresponding intersection circles with the boundary.

Lemma 5.4.1: The problem of 3D border coverage of a cube by sensor nodes modeled as 3D balls is equivalent to the problem of complete coverage of a 2D plane by sensor nodes modeled as circles.

Proof: According to the definition of border coverage, each point on the border should be covered by at least one sensor. The border $B(R)$ of the cubical region R is represented by 6 faces (2D planes). First, if each face of the cube $B_a \in B(R)$, $a=1,2,3,4,5$ *and* 6 is completely covered by a set of circles $Cir_a = \{Cir_{a1},..Cir_{an}\}$ and if D_i is the disc bounded by the circle Cir_i then $\forall p \in B_a, p \in D_{ai}$ *for some* $Cir_{ai} \in Cir_a$. The 3D border coverage is now transformed to finding the spheres whose border intersections are these circles. $\forall p \in B(R), p \in A_i$ *for some* $Cir_i = A_i \cap B(R)$. Now, if we have a set of sensor nodes that border cover a 3D cubical region, taking the intersection of the spherical sensing regions of the sensor nodes with each face of the cube will result in the formation of circles which completely cover the 2D plane. So, the 3D border coverage problem is transformed to the 2D full coverage problem. □

Let the intersection of any boundary plane B and sphere S_i be circle Cir_i, i.e. $Cir_i = B \cap A_i$. The interior of the circle Cir_i is said to be the disc bounded by the circle Cir_i, i.e. $D_i = interior(Cir_i)$.

Definition 5.4.1: A circle Cir_i is completely covered if the disc bounded by the circle is completely covered, i.e. $\forall p \in D_i, p \in \bigcup\limits_{j=1}^{n} A_j.$ □

Definition 5.4.2: A sensor S_i is a **border-redundant** sensor if $Cir_i = B \cap A_i$ is completely covered by neighboring spheres. □

Lemma 5.4.2: A sensor S_0 is border redundant if all the intersection points $Cir_i \cap Cir_j \in D_0, \forall i, j = 1...n$ are covered by one or more adjacent sensor nodes.

Lemma 5.4.2 indicates that a sensor node S_0 is border redundant if all the intersection points $Cir_i \cap Cir_j \in D_0$ are covered by some sensor $S_l, l \neq i, j = 1..n$. Therefore, to check if S_0 is border redundant; one has to first find all the circles obtained by the intersection of $S_0 \cap B_m, m = 1..6$. For each Cir_k, find all the intersection points that lie within D_k. If all these intersection points are covered, then the circles Cir_k are covered. Then, by the theorem 4.2.2, S_0 is border redundant and can be deactivated without affecting the overall cubical border coverage.

3D Distributed Border Coverage Algorithm

For each node S_i, form the set of neighbors $N(i)$.

Step 1: _Find the intersection circles_ Cir_i

Find the intersection circle Cir_i resulting from the intersection of S_i's sensing region with the boundary of the region of interest. Note: S_i might intersect 2 or 3 boundary planes of the region in semi circles. The same procedure will still apply.

Step 2:

For every pair of nodes S_k, S_l in $N(i)$

- Find the intersection circle $Cir_{k,m} = A_k \cap B_m$ and $Cir_{l,m} = A_l \cap B_m$

  (*where B_m is the boundary plane being tested*)

- Find the intersection points $Cir_{k,m} \cap Cir_{l,m}$

- If the intersection points are all covered, i.e. $Cir_{k,m} \cap Cir_{l,m} \in A_n$, $S_n \in N(i)$, $n \neq i,k,l$, then deactivate S_i since it is a border redundant node.

The algorithm guarantees that every point on the boundary of the target region is covered by at least one sensor. The minimal set of sensor nodes is also selected. The computational complexity of the algorithm developed in this section is $O(n.N^3)$ where n is the total number of sensor nodes in the network and $N = \left(\max_{i=1}^{n} |N(i)| \right)$.

The key to both 2D and 3D border coverage algorithms is that they are performed in distributed manner. The 3D distributed coverage algorithm requires that each sensor node knows the information about locations of all sensing neighbors. The algorithm maintains a table of known sensing neighbors based on the beacons (HELLO messages)

145

that it receives from its communication neighbors. Assuming that $R_c>2R_s$, the sensor nodes need to include only their locations in the HELLO messages. When a network is deployed, all nodes are initially active. Redundant nodes will switch to the inactive mode until no more nodes can be turned off without causing coverage holes in the region. The distributed algorithm consists of two steps. First, each node advertises its position and listens to HELLO messages from other nodes to obtain neighboring nodes' position information. Secondly, each node runs the border coverage algorithm (2D or 3D) discussed earlier and decides whether to deactivate or not. The details of these two steps are introduced as follows. To obtain neighbor node information, a simple approach is that each node broadcasts a HELLO message, which contains node ID and its current location, at the beginning of each round. Note: If nodes have different sensing ranges (due to depletion of power), the message should also include the current sensing range of the transmitter as well. After finishing the collection of neighbor information, each node evaluates its eligibility for turning off by running the 3D coverage algorithm. However, if all nodes make decisions simultaneously, blind points may appear. To avoid such a problem, each node announces to its neighbors that it is currently running the coverage algorithm. If the node is redundant and is eligible for turning off without affecting the overall coverage, it will broadcast a GOODBYE message to its neighboring nodes. Neighboring nodes receiving a GOODBYE message will delete the sender's information from their neighbor lists.

Since the algorithm does not produce a unique result, it is advantageous to have a performance measure for comparing two different collections of sensor nodes that border cover a region. The measure of optimality of a border cover of a WSN is the ratio of the number of active border sensor nodes in the network to the minimum number of sensor nodes that can border cover the same region. The results in section 5.3 found the locations of sensor nodes to achieve optimum deployment for border coverage a region R. Therefore, given the region to be monitored for border coverage, one could easily find the number of sensor nodes required and their location for border coverage.

5.4.3 Optimizations of the Distributed Algorithm

In this section, an optimization to the border coverage selection algorithm is provided in order to improve the border coverage life time of the region. Sensor networks introduce new challenges for fault-tolerance. The algorithms presented in section 5.4.1 allow us to select a minimum subset of the sensor nodes already deployed that will guarantee border coverage of a given region. However, the emergence of border coverage holes in the target area is unavoidable

A proactive method of utilizing the total energy is to assign tasks for each sensor nodes so that a hole is never formed. Though this solution might give optimal solutions, it is impractical in real time applications. In this section, we provide a reactive and practical approach to minimizing border coverage holes as they (or before) are formed. We provide this self healing algorithm as an enhancement to the border coverage algorithm developed in the previous section. We call it "self heal" as the actuation is not governed by a user command or application but initiated by the WSN to salvage its own performance.

We assume that the nodes know their initial energy content and can keep track of their energy expenditure and therefore can predict their own death. Sensor nodes are randomly deployed in a region of interest to be border covered. Every node acquires information about its location and communication neighbors. The border coverage selection algorithm developed in section 5.4.1 is run in a distributed fashion on each node and an optimal border cover of the region of interest is selected. However, while running the border coverage selection algorithm, each sensor node keeps track of the border neighbors i.e. the neighboring nodes who are also border sensor nodes. If a node is about to run out of energy (before the energy level goes below a specific threshold), it runs the selection algorithm on its border neighbors to select an optimal set of sensor nodes to be its substitute border cover i.e. to cover its border intersection in case of its death. It then broadcasts a HELP message in order to activate the sleeping nodes that will minimize the border coverage hole.

Sensor nodes can be also misplaced or destroyed accidentally or deliberately. Since each node knows its own location and whether it is a part of the border cover set or not, upon realizing the malfunctioning of its sensor, a node broadcasts a HELP message in order to cover the border coverage hole. This simple extension of the border coverage algorithm results in better energy utilization and extends the border coverage life time of the region.

5.5 Identifying Coverage Boundary

Several things can occur in the wireless sensor network that can impair their functionality. The target field that is supposed to be completely covered by the densely deployed nodes may have coverage holes, i.e. there might be areas that are not covered by any node *(coverage hole problem)*. The network fails to achieve its objectives if some of the nodes cannot sense or report the sensed data. Some of these anomalies may be deliberately created by adversaries that are trying to avoid the sensor network.

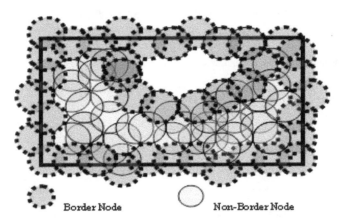

Figure 5.7(a): A region with some coverage holes and the sensor nodes that lie on the boundary of the coverage and coverage holes.

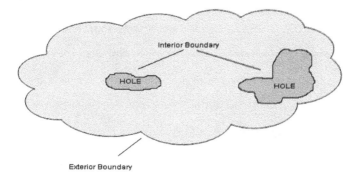

Figure 5.7(b): Exterior Boundary (Line surrounding the yellow interior) and Interior Boundary (Line surrounding the gray interior) illustration.

In this section, a distributed algorithm to solve the *"coverage boundary problem"* is devised i.e. detect the find those sensor nodes that are on the boundary of coverage (or coverage holes) in the region. Assuming that the boundary of the region is known, we can easily discard those nodes that are on the boundary of the region (if required). The task of identifying the sensor nodes that lie on the exterior or interior boundary would be much easier if it was done in a centralized way. A single node (with no energy constrains) is aware of the exact location of all the sensor nodes in the region i.e. it can form an image of the sensor distribution. Edge detection techniques in image processing provide an automatic way of finding boundaries of one or more objects in an image. For each pixel in the image we measure the color intensity of the pixel and subtract the color intensity of nearby pixels. If the pixel lies in a region with sharp changes in intensity then the intensity difference will be large thus indicating a boundary edge and the corresponding sensor node on the boundary could be easily identified. However, in real time, distributed and localized algorithms need to be adapted due to the energy constrains of a single sensor node and centralized algorithms are considered impractical.

Let the set of all sensors that lie either on the boundary of coverage or on the boundary of some coverage hole H in **R** be denoted by $C_{boundary}$. A boundary sensor is therefore any sensor that is in $C_{boundary}$ i.e. $S_i \in C_{boundary} \Leftrightarrow \exists p \in Cir_i \mid p \notin A_j, \forall S_j \in S$.

The work in this chapter is a continuation of our previous work where we developed algorithms to deactivate redundant sensors in the region and end up with a minimal set of sensors that fully covers the region of interest. Let the final reduced set be denoted by C_{full}. We will start by first proving that if we had already found C_{full}, the reduced cover of a region, then $C_{boundary}$, the set of sensors that lie on the boundary of the region or holes in the region would be the same had we started with the original set of sensors. This theorem is very useful since the whole goal is to minimize the energy consumption by reducing the number of active nodes and depending on the application, we could either activate the reduced cover set or a subset of it ($C_{boundary}$) which will further minimize the energy consumption.

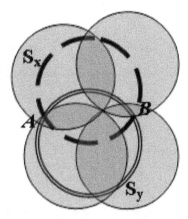

Figure 5.8: S_x (dashed circle) is a redundant sensor. Deactivating it won't transform its intersecting non boundary sensor S_y (double line circle) to a boundary sensor node.

Theorem 5.5.1: The set of sensors that lie on the coverage boundary is always a proper subset of the reduced cover set i.e. $C_{boundary} \subset C_{full}$.

Proof: Suppose we have a redundant sensor s_x (Figure 5.8) which will be deactivated by using any reduced cover algorithm. If we can prove that by deactivating s_x, none of the remaining non-boundary sensors will be transformed to a boundary node, then the final reduced cover set of the given region will always include $c_{boundary}$. Suppose s_x intersects sensors s_y, s_z, s_w ... Let s_y, s_z, s_w ... be non-boundary nodes before s_x is deactivated. Since s_x is redundant, then the intersecting regions of s_x and s_y, s_z, s_w ... are clearly covered by other sensors. In Figure 5.8, consider the intersecting Arc AB inside s_x (covered by s_x).

By deactivating s_x, the interior arc AB is still covered by s_x's neighboring nodes. According to the definition of a non-boundary sensor, all the points that are on the boundary circle are covered and by deactivating s_x, all the points on the boundary circle are still covered. So, none of the non-boundary sensor nodes will be transformed to boundary nodes which implies that $C_{boundary} \subset C_{full}$. □

Definition 5.5.1: An arc is any portion of a circle of any given angle. It is represented by 2 points, start point and end point (in degrees) going counter clock wise. □

Definition 5.5.2: A sensor S of circumcircle (sensor boundary) *Cir* is said to be **boundary covered** by a set of arcs $S_c = \{Arc_1, Arc_2, ..., Arc_n\} \Leftrightarrow \bigcup_{i=1}^{n} Arc_i = Cir$. □

Based on the definition of a boundary sensor the selection algorithm will look at the boundary (circumference) of each sensing circle and determine whether it is completely covered or not. If the circumference of the sensor's circle is completely covered by its neighbors then it is not a boundary sensor and therefore could be deactivated without affecting the overall boundary coverage.

Consider 2 sensor nodes: s_1 of center $O_1(x_1, y_1)$ and s_2 of center $O_2(x_2, y_2)$ with equal sensing radii $R_1 = R_2 = R$. We want to inspect what part of s_2 is boundary covered by s_1.

The distance d between the 2 sensors is easily calculated as $d = \sqrt{(x_1 - x_2)^2 + (y_1 - y_2)^2}$. In

Figure 5.9, the portion of circle C_1 that is covered by C_2 is arc AB which can be

calculated from angle $\angle AO_2B$. In triangle $A\overset{\Delta}{O_1}O_2$, $R^2 = R^2 + d^2 - 2.R.d.\cos(x) \Rightarrow x = \arccos(\dfrac{d}{2r})$. So

the portion of C_2 that is boundary covered by C_1 is the arc $[180 - x, 180 + x]$.

Given a sensor s_i being tested for boundary coverage, any neighboring sensor s_r is

set as the reference neighbor and the portions covered but the other neighboring nodes

(for example s_j) is calculated in terms of the reference axis by a simple rotation

clockwise by angle $o_r\overset{\wedge}{o_i}o_j$. In Figure 5.10, the portion of C that is covered by C_1 is arc

$AB = [180 - z, 180 + z]$ referring to $Axis_1$. C_1 is set as a reference center and the corresponding

reference axis $Axis_1$ (line containing O_1O , line perpendicular to O_1O). The potion of C

covered by C_2 is therefore $CD = [180 - w, 180 + w]$ on a different axis ($Axis_2$). Referring to the

same reference axis ($Axis_1$), $CD = [180 - w - x, 180 + w - x]$ where $x = \angle O_1OO_2$. For C_1 to be

boundary covered, the total covered portion should be $[0, 360]$.

Definition 5.5.3: We call Arc_j the **successor** of Arc_i denoted by: $Arc_j = \zeta(Arc_i)$ if all the

following conditions are satisfied:

➢ $Arc_j \cap Arc_i = Arc_{ji} \neq \varnothing$

➢ $x_j > x_i$

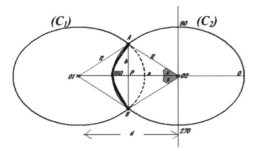

Figure 5.9: Two sensors S_1 and S_2 intersecting. The portion of C_2 that is covered by C_1 is arc AB.

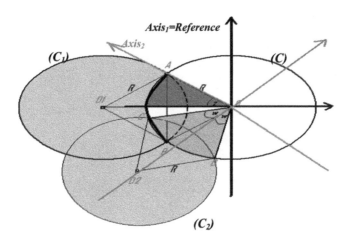

Figure 5.10: Sensor S of circum circle C is partly boundary covered by sensors S_1 and S_2.

In Figure 5.10, Arc CD is the successor of Arc AB i.e. $Arc(CD) = \zeta(Arc(AB))$. Note that the ending point of Arc_i may coincide with the starting point of Arc_j.

Definition 5.5.4: Arc_i is said to be **contained** in Arc_j ($Arc_i \subset Arc_j$) if the endpoints appear in counter clockwise order as x_j, x_i, y_i, and y_j (Figure 5.10)   □

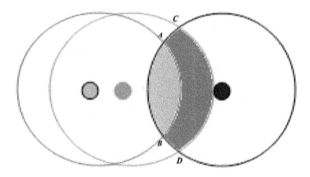

Figure 5.11: Example of arc containment. $Arc_i (= AB) \subset Arc_j (= CD)$.

The first step in the algorithm is to remove those arcs that are contained in another arc and thus end up with a set of arcs with no arc contained inside another. Using definitions

5.5.1-5.5.4, the following theorem will be proven and would be the key to the distributed algorithm.

Theorem 5.5.2: A set $S_c = \{Arc_1, Arc_2, ..., Arc_n\}$ of n circular arcs (with no arc contained in another arc) do not completely cover (the circumference) C **if and only if** an arc $Arc_i \in S_c$ for some i, has no successor i.e. $\exists Arc_i \in S_c \mid \zeta(Arc_i) = \varnothing$.

Proof: We prove the "*if*" part by contradiction. Suppose $S_c = \{Arc_1, Arc_2, ..., Arc_n\}$ does not cover C completely and there is an arc $Arc_x \in C$ that is not covered by S_c. Suppose all the arcs in S_c have successors i.e. $\forall Arc_i \in S_c, \zeta(Arc_i) \neq \varnothing$. Without loss of generality, suppose that the arcs are ordered as $Arc_1, Arc_2, ..., Arc_n \mid Arc_2 = \zeta(Arc_1), ... Arc_n = \zeta(Arc_{n-1})$

Since all the arcs have successors then $Arc_1 = \zeta(Arc_n)$ which implies that the arcs made a complete (360^0) cover on C. Therefore Arc_x should be covered by $S_c = \{Arc_1, Arc_2, ..., Arc_n\}$ .Thus, we reached a contradiction and therefore all the arcs in S_c can't have a successor i.e. $\exists Arc_i \in S_c \mid \zeta(Arc_i) = \varnothing$.

To show the "*only if*" part, suppose $\exists Arc_a \in S_c \mid \zeta(Arc_a) = \varnothing$. Let us study the three conditions that might result in Arc_a having no successor.

1. $\forall Arc_b \in S_c, Arc_b \cap Arc_a = \varnothing$

2. $x_b < x_a$

3. $\exists$ starting point $\in A_{ba}$.

Suppose condition 1 is true leading to Arc_a having no successor. It means that Arc_a is independent. Since there are no arcs contained in another arc that means that Arc_a will result in uncovered part of C. Suppose condition 1 is not true i.e. $\exists Arc_b \in S_c \mid Arc_b \cap Arc_a = Arc_{ba} \neq \varnothing$ and that condition 2 is true for all arcs Arc_b intersecting Arc_a. Since there are no starting point of an intersecting arc that goes beyond Arc_a's starting point, then we can't make a whole rotation on C and thus C is not completely covered.

Suppose conditions 1 and 2 are invalid i.e. $\exists Arc_b \in S_c \mid Arc_b \cap Arc_a = Arc_{ba} \neq \varnothing$ and $x_b > x_a$ but condition 3 is true. Since there is a starting point on the intersection arc and conditions 1 and 2 are true then Arc_a must have a successor but this is a contradiction to the given.

After analyzing all the possible cases, we conclude that if $\exists Arc_i \in S_c \mid \zeta(Arc_i) = \varnothing$ then C is not completely boundary covered by S_c. $\square$

Using theorem 5.5.2, a computational efficient distributed algorithm that will select the sensor nodes that lie on the boundary of the coverage or coverage holes in the region can now be developed. The algorithm starts with an initialization phase where each node forms its neighbor set and then evaluates whether or not it lies on the boundary of a coverage hole depending on its boundary coverage by its neighbors.

Distributed Coverage Boundary Algorithm

For each node S_i, form the set of neighbors $N(i)$.

The covered set of arcs in Cir_i, SA_i, is initially set to $\varnothing$

Step 1: Containment Elimination and Arc set Formation

For every pair of nodes S_j, S_k in $N(i)$

- Find the portion of Cir_i, $Arc_j = [x_j, y_j]$, and $Arc_k = [x_k, y_k]$, which are arcs covered by S_j and S_k using the procedure discussed earlier.

- Check if the endpoints appear in clockwise order as x_j, x_k, y_k, and y_j.

If this condition is satisfied then $Arc_k = Cir_k \cap Cir_i$ is contained in $Arc_j = Cir_j \cap Cir_i$ and can be ignored.

- Update the neighbor set of S_i to ignore the sensors with redundant arcs i.e. $N(i) = N(i) - S_k$.

- $SA_i = SA_i \cup \{Arc_j\}$.

Go to step 2.

Step 2: Successor Testing

Successor set ζ is initially set to SA_i.

For each element $Arc_m = (x_m, y_m)$ in SA_i do the following:

For each element in $Arc_n = (x_n, y_n)$ in $SA_i - Arc_m$ check if $x_m \leq x_n \leq y_m \leq y_n$. If the condition is true then for each element $Arc_k = (x_k, y_k)$ in $SA_i - Arc_m - Arc_n$ check if $x_n \leq x_k \leq y_m$. If that condition is False for each element then $Arc_n = \zeta(Arc_m)$ and $\zeta = \zeta - Arc_m$.

If after performing steps 1 and 2, if $\zeta = \varnothing$ then every arc has a successor and therefore the circumcircle of the given sensor node is completely covered and is declared as a non-boundary sensor node.

The computational complexity of the redundancy algorithm developed in this section is $O(n.N^3)$ where $N = \left(\max_{i=1}^{n} |N(i)| \right)$ is the maximum number of nodes in the neighbor set of any sensor in the network. Assuming we have a large network of size n, the computational complexity will be $O(n) | n \gg N$.

5.6 Border Patrol Strategy

Surveillance has been a typical application of wireless sensor networks. To conduct surveillance of a given area in real life, one can use stationary watch towers, or can also use patrolling sentinels. Comparing them to solutions in sensor network surveillance, all current coverage based methods fall into the first category. In the previous sections, we minimized the number of active sensor nodes in order to detect a target intrusion at all time. In this section, we consider sleep scheduling of the border sensor nodes. A surveillance sensor network is desired to operate unattended for a long time, usually much longer than the battery life-time of a single node. Thus, power conservation is critical and over deployment of sensor nodes is necessary. Each border node can swap between working and sleeping modes and the network only maintains a subset of working nodes. Moreover, sleep scheduling plays an important role in sensor placement planning. It is very important for the network to let each node have longer sleep time, however, still maintain certain level of ability in detection.

We propose a *Border Perambulation* model for surveillance operations in sensor networks. The goal is to put the sensor nodes into deeper sleep mode and still detect any interesting even at all times. In the region of interest, at each point of time, only a very small subset of the border nodes are active (possible one border node) for intruder detection. As the time progresses, the active nodes move along the border, so that it will sweep the entire border of the region. This procedure of sweeping coverage can be repeated with a given period. Under this *Border Perambulation* model, the network's power consumption rate is much lower than the conventional surveillance operations. However, this method has to provide ensured target intrusion detection of the region at all time. We will present the required conditions to guarantee target detection while using the border patrol technique. The *Border Perambulation* problem is studied as two separate cases:

Case1: In the first case (Figure 5.12), we suppose that the time for one whole loop on the sensors on the border cover is **variable** and depends on the maximum speed of the intruder. Let V_{max} be the maximum speed of intruder and R_s be the sensing radius of each sensor node. The time of one whole loop should be $t = \dfrac{2R_s}{V_{max}}$ in order to guarantee the detection of an intruder at all times. Knowing the total time of one whole loop on the border sensor nodes, the sleeping schedule of each border node can be calculated. Suppose we have n border nodes then the wake up time of each border sensor node would be $T_{wakeup} = \dfrac{2R_s}{n.V_{max}}$. Since each node is aware of its location and the location of its neighbors, we could select any border node as an origin border node $S_{Border0}$ and move clockwise to label the border nodes $S_{Border1}$, $S_{Border2}$... The border patrol then will start with the dissemination of the patrol setup containing the wake up time of each sensor.

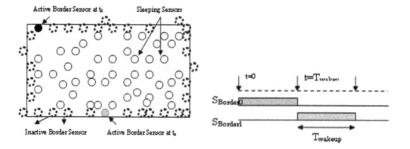

Figure 5.12: (a) Example of a border patrol (case 1) and its corresponding wake-sleep schedule.

162

Case 2: Another case to be considered is when the optimal border deployment of the sensor nodes is established as shown in Figure 5.13(a). The nodes are deployed optimally one level at a time. The first set of border nodes (level 1) are deployed using an optimal deployment procedure discussed earlier. The center of each sensing region is on the border edges. The borders nodes of level 2 are obtained by optimal border deployment of the nodes on an "Imaginary Rectangular region" of dimensions $L - R_s, W - R_s$ where L and W are the dimensions of the region of interest (see Figure 5.13(a)).

In this case the border patrol algorithm assumes that the maximum speed of the target is known and the time of the iteration is constant. We want to determine the number of levels of border coverage needed to guarantee detection of the target. Let t be the time for one cycle of the patrol. Let V_{max} be the maximum speed of intruder and δ be the level of border coverage needed for detection of an intruder at all time then $\delta = \dfrac{V_{max} \cdot t}{R_s}$ is the required level of border coverage. So, we could know exactly how many levels of border coverage need to be activated in order to detect an intruder at all times.

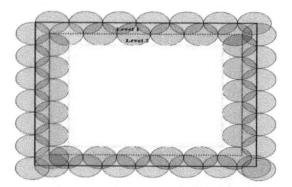

Figure 5.13(a): Example of an optimal deployment with multiple levels.

<u>Case 3:</u> In the third case (Figure 5.13(b)), a constant iteration time 't' is assumed and the thickness of the border cover and the corresponding optimal wake up time of each sensor node to guarantee detection of the target at all time is required. The thickness of the border cover is the size of the active patrol zone which is a square region of side θ. Let V_{max} be the maximum speed of intruder and θ be the thickness of the border cover. Then, the maximum distance traveled by the target would be $V_{max} t$. So, the minimum thickness of the cover in order for the target to be detected at all time is $\theta = V_{max} t$. Since the size of the active zone in order to guarantee target detection is known, algorithms developed in previous sections could be used to find a reduce cover of the active zone and the sleep schedule of each active zone is determined (unlike the first case, where the sleep schedule of each individual border node was to be determined). Suppose that the region to be

monitored is a rectangular plane of length L and width W. Then the wake up time of each

border sensor is the total time divided by the number of active zones i.e.

$$T_{wakeup} = \frac{t}{\left\lceil \dfrac{2(L+W)}{\theta} \right\rceil} = \frac{t}{\left\lceil \dfrac{2(L+W)}{V_{max} \cdot t} \right\rceil}$$ where $\lceil a \rceil$ is the smallest integer greater or equal to a.

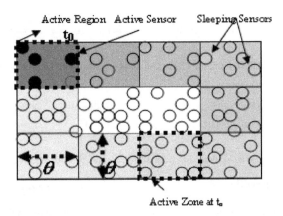

Figure 5.13(b): Example of a border patrol (Case 3).

5.7 Simulation Results

The theoretical developments in Sections 5.2-5.6 are validated through numerical

examples in this section. The case of random deployment of sensor nodes is studied and

compared to the optimum deployment for border coverage. Both 2D and 3D cases are

considered and the number of sensor nodes required for border coverage is studied. The number of sensor nodes required to cover a 2D region of size 10 units by 10 units (or a 3D region of size 10x10x10) is considered. Random deployment, optimal deployment and optimal selection of the nodes for border coverage are studied for different values of the sensing radius. The optimization to the border algorithm is also tested and resulting border coverage lifetime of the network is analyzed.

To test for border coverage, the region of interest is divided to a 2D or 3D grid and a centralized algorithm is developed to test for border coverage by generating an occupancy grid and checking if the first and last row and the first and last column in this grid are covered. If all the cells in the first and last row and the first and last column are occupied, then the entire region is border covered. The region to be covered is divided into squares of side equaling half the sensing radius of each sensor nodes. Since the region to be covered is divided into a grid with cell size equal to $\frac{R_s}{2}$, any cell in this grid is completely covered only if its center is within a distance of $\frac{R_s}{2}$ from the sensor. Since we are only concerned with the border coverage, as can be seen from Figure 5.14, at most 12 cells need be checked to verify the border coverage of a sensor and a maximum of $12n$ cells need to be checked for the border coverage region of 'n' sensor nodes.

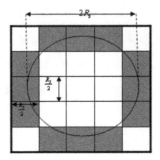 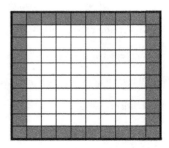

Figure 5.14: The coverage region of a sensor node (in2D) and the occupancy grid of n sensor nodes with highlighted border.

In the 1st experiment, the optimum 2D coverage algorithm is used to find the optimum border cover of region 10x10 units when sensor nodes are randomly deployed. The nodes have a sensing radius of 1 unit and initially different numbers of nodes are randomly deployed in this region using a uniform distribution. It can be seen that the average *optimality measure* of the border selection algorithm is 1.228 and the nodes that were active in the optimum border cover resulted in average savings of 98.4% (when the number of deployed nodes 500,1000,1500,2000,2500, and 3000) (Figure 5.15(a)). In Figure 5.15(b), the required number of sensor nodes with different radii using random deployment, optimal 2D Border deployment and 2D Border selection algorithm are compared.

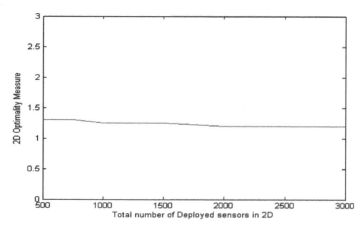

Figure 5.15(a): The optimality measure of the border selection algorithm for different number of deployed nodes in 2D.

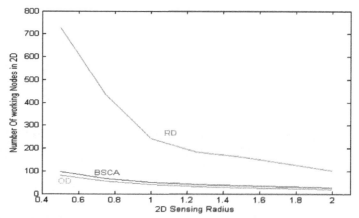

Figure 5.15(b): Comparison between RD (Random Deployment), OD (Optimal Deployment), and BSCA (Border Selection Cover Algorithm) in 2D.

In the 2nd experiment, the same comparison (Figure 5.16(a)) is done for the 3D case and the resulting average optimality measure is 1.123 and when the 1500, 2000, 2500, 3000, 3500, and 4000 nodes were randomly deployed, the border selection algorithm resulted in average savings of 93.71%. In Figure 5.16(b) the required number of sensor nodes with different radii using random deployment, optimal 3D Border deployment and 3D Border selection algorithm are compared.

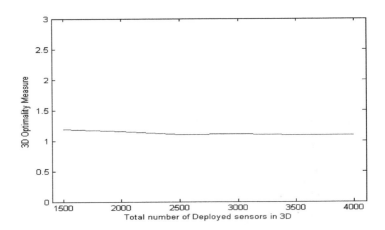

Figure 5.16(a): The optimality measure of the border selection algorithm for different number of deployed nodes in 3D.

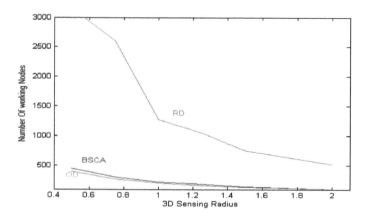

Figure 5.16(b): Comparison between RD (Random Deployment), OD (Optimal Deployment), and BSCA (Border Selection Cover Algorithm) in 3D.

In the 3rd experiment, we evaluate the border coverage percentage of the region when the sensor nodes are randomly deployed and the border coverage selection algorithm is applied. As we vary the number of deployed nodes, the border coverage of the region using the border cover obtained (Figures 5.17(a) and 5.17(b)) is evaluated. It is noticed that after a specific threshold value for 2D and 3D cases, the border coverage percentage is always one. The reason is that random deployment of the sensor nodes does not guarantee border coverage of the region below that threshold.

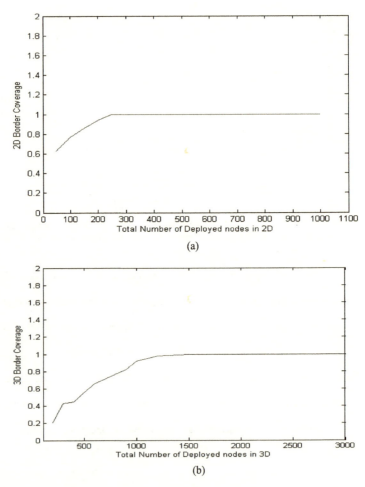

Figure 5.17: (a), (b) Varying the number of deployed nodes will result in different border coverage percentage for both 2D and 3D regions of interest.

In the 4th experiment, the system life time is tested. The metrics used in evaluating system lifetime is the *border coverage lifetime*. The overall border coverage lifetime is the continuous operational time of the system before the border coverage drops below its specified threshold (for example 0.9). In the Figures 5.18(a) and 5.18(b), the system lifetime is evaluated assuming that each sensor node has a limited energy supply (300 Joules) and when it runs out of energy it is deactivated. The node deployment densities are 300 and 600 respectively. We started with 300 nodes deployed since that is the minimum number of nodes that will guarantee border coverage of the region using random deployment. With each density, the nodes are randomly distributed in a 10×10 region network field and each of them starts with an initial energy of 300 J. The power consumption of Tx (transmit), Rx (receive), Idle and Sleeping modes are 1400mW, 1000mW, 830mW, 130mW respectively. As time passes, sensor nodes will be deactivated due to lack of energy and will leave some coverage holes in the border of the region. If 300 sensor nodes were deployed, after approximately 1600 seconds, the border coverage percentage using the original network will drop below 0.9. However, using the border selection algorithm it needs about 2300 seconds to drop below the threshold. If we increase the number of deployed nodes to 600, the cost for calculating the border cover will increase and thus after approximately 1690 seconds the border coverage percentage will go below 0.9. In both experiments, the border coverage life time of the network using the border selection algorithm is much better that that using the original network. In Figure 5.18(c), we divide the border of the region into 1000 grid points and test how many sensor nodes cover each grid point before and after running the algorithm. When the number of deployed nodes is 600 , we could see that before starting the algorithm, the degree of border coverage is much higher that that after running the algorithm which implies that the random deployment is not optimum and therefore a lot of energy is wasted due to multiple active nodes in a given border region. After running the algorithm, most of the redundant border nodes are deactivated resulting in a much energy efficient deployment of the nodes.

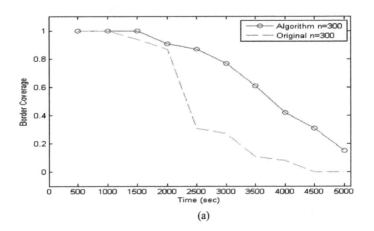

(a)

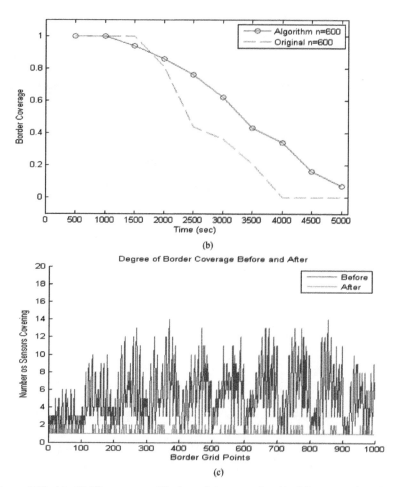

Figure 5.18: (a), (b) The coverage life time of the network with different number of deployed sensor nodes. (c) The degree of border coverage before and after running the border selection algorithm.

In the 5th experiment, we do the same comparison that was done in the 4th experiment however with the optimizations mentioned in section 5.4.3 added. We notice that the system life time (border coverage life time) is much better than the case if we had started with the original set of deployed nodes. The strength of the developed algorithm is that it allows the sensor network to adaptively reconfigure and repair itself in order to improve its own performance. In Figure 5.19, as we increase the number of deployed nodes (from 300 to 600 nodes), the self healing border coverage algorithm performs better since activating a substitute set will result in better percentage of border coverage and therefore the border coverage lifetime of the network is increased.

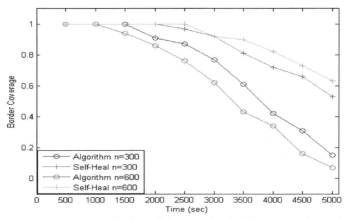

Figure 5.19: The coverage life time of the network with different number of deployed sensor nodes when using self healing enhancement of the algorithm.

In Figure 5.20, an example of the active nodes before and after running the algorithm is presented. 2000 nodes were deployed, and after running the border selection algorithm, 1974 nodes were deactivated resulting in savings of 98.7%.

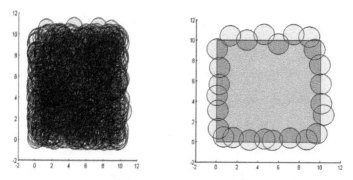

Figure 5.20: An example of the border selection algorithm. Active nodes, before and after running the algorithm are shown.

Minimizing the number of sensor nodes active to border cover a region of interest will result in minimizing the energy consumed by the whole sensor network and thus increasing the life time of the network as demonstrated in the simulation results. The theoretical developments in sections 5.5 and 5.6 are validated through the following experiments. Given a random deployment of sensor nodes, the distributed hole coverage algorithm is applied and the sensor nodes that lie on the boundary of the coverage holes will broadcast a message (or will stay active while all other nodes deactivate themselves). The distributed hole coverage algorithm described in section 5.5 is used to find the hole boundary covers in a region of 10x10 units when sensor nodes are randomly deployed. The nodes have a sensing radius of 0.75 unit and initially 250 nodes are randomly deployed in this region using a uniform distribution. Figure 5.21(a) shows the initial deployment of the sensors and Figure 5.21(b) selects the sensor nodes that are on the boundary of some coverage holes in the region using the distributed algorithm. It can be seen that 31 nodes were on the boundary of 6 coverage holes in the region. Figures 5.22 and 5.23 show that as the number of deployed nodes increase (or sensing radius increase), the number of sensors on the boundary of the holes decrease until it reaches zero when there are enough sensor nodes to completely cover the region of interest. Note that we only consider "bounded" coverage holes i.e. holes that are surrounded by active sensor nodes (not by sensor nodes and the boundary of the region). We also discard nodes that are on the boundary of the region since the algorithm will detect both nodes on the boundary of the coverage and nodes on the boundary of the coverage holes.

In the last experiment, we compare the distributed hole coverage boundary algorithm to the centralized algorithm. Note that the centralized algorithm identifies the holes in the region and the size of the hole. However, it doesn't identify the nodes that are on the boundary of the holes. To do that, we adjusted the centralized implementation algorithm to first select the boundary grid points that lie on the boundary of each coverage hole (0 entries) and then estimate the sensor nodes that lie on the boundary of the hole by minimizing the perpendicular distance from the boundary grid points to the sensing region of each close sensor. Figure 5.24 depicts excellent performance of the distributed algorithm relative to the centralized version. The ratio of the number of holes calculated using the distributed algorithm to that using the centralized version always remains close to the ideal value of 1. It is not always equal to 1 since the algorithm considers only bounded holes. However, the centralized hole detection algorithm considers all holes in the region of interest. The simulation results show that the distributed algorithm performs very well compared to the centralized algorithm.

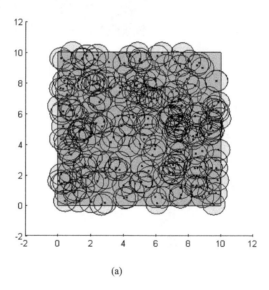

(a)

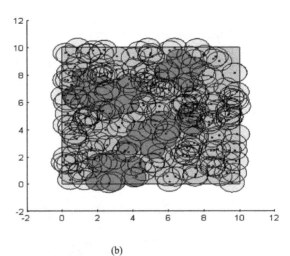

(b)

Figure 5.21: (a) 250 sensor nodes with sensing radius 0.75 units are randomly deployed in a 10x10 region. (b) 31 sensor nodes are on the boundary cover of the 6 bounded coverage hole.

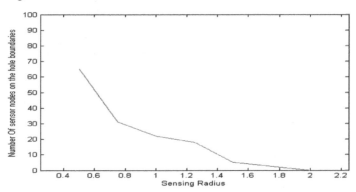

Figure 5.22: Different sensing radii of sensor nodes in a 10x10 region compared to the number of sensor nodes on the boundary of the bounded coverage holes.

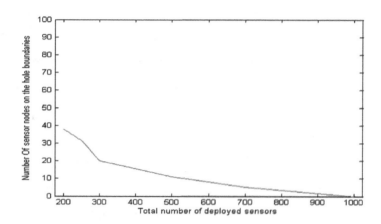

Figure 5.23: The number of deployed nodes of sensing radius 0.75 compared with the number of sensor nodes on the boundary of the coverage holes in a 10x10 region.

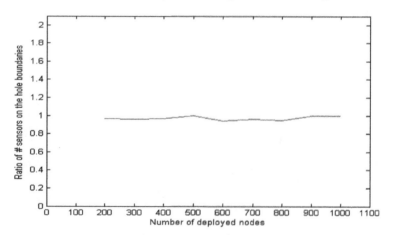

Figure 5.24: The ratio of the number of sensor nodes calculated by the distributed algorithm to that calculated by the centralized algorithm.

In the next experiment, we evaluate the system life time. We compare between 4 techniques. The border coverage life time using the original network, a border cover, a border patrol cover with minimum speed for detection, and a border cover patrol with double the minimum speed. As shown in Figure 5.25, we could see that the border patrol with minimum speed outperforms all the others. The border coverage life time is maximized. When doubling the speed of the virtual patrol, the border coverage life time decreases, since nodes will wake up twice as much had we used the minimum speed for target detection. However, as shown in Figure 5.26, the delay of detection will decrease linearly as we increase the speed of the patrol. So, it is trade off between the coverage lifetime and the delay of detection depending on the application under investigation.

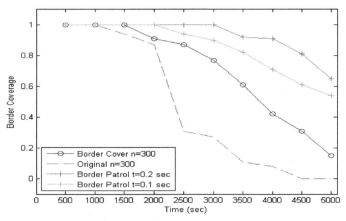

Figure 5.25: The system lifetime as time passes.

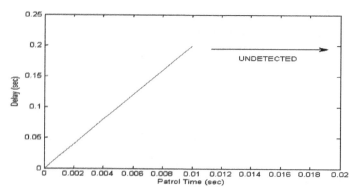

Figure 5.26: Varying the speed of the border patrol will result in smaller delay of detection.

5.8 Chapter Conclusions

In this chapter, two different problems: the boundary coverage and the coverage boundary problems were analyzed. Algorithms were proposed to compute the minimum number of sensors required for boundary coverage of a given region. We also provided algorithms which not only find the exterior coverage boundary of a network but also identify the boundary of the coverage holes in the network. The coverage problem (Full Coverage, Boundary Coverage) is one of the fundamental problems in sensor networks and can be used to provide energy efficient utilization of the sensors. A sensor network for detecting large scale phenomena such as containment flow or seismic disturbance may be used to describe the boundary of the phenomenon. In such cases, it is crucial to identify the sensor nodes that lie on the boundary of the coverage region. Part of the

182

future work is to use the algorithms developed in this chapter for tracking applications. The algorithms presented can be easily extended to handle different shapes of region to be monitored. If the region of interest is of an irregular shape, we can always use polygon approximation and simplification techniques to find the polygon that bounds the region of interest. In addition to that, the sensing radius of each sensor node need not be equal and the distributed algorithm could be applied to sensor networks with different sensing radii.

In the next chapter, algorithms developed in chapter 3-5 will be used to provide and energy efficient tracking algorithm using WSNs.

Chapter 6

Tracking using WSNs

Distributed algorithms for boundary coverage were provided in the previous chapter. In this chapter, a novel approach for tracking a dynamic phenomenon is presented. Algorithms from Chapters 3, 4, and 5 will be used in order to implement an energy efficient distributed tracking algorithm using WSNs. One of the central issues in sensor networks is energy efficient target tracking, where the goal is to monitor the path of a moving target using a minimum subset of sensor nodes while meeting the specified quality of service (QoS). Unlike other tracking methods that are based on computationally complex clustering techniques, the strategy is based on finding a reduced cover of the whole region and then subdividing the reduced cover into sub covers based on the target's location. The behavior of the proposed tracking algorithm is analyzed through simulation and the excellent performance is illustrated. We study the tradeoff involved in the energy efficient tracking of the target and compare the performance of the distributed tracking algorithms with other popular strategies. The gain in energy savings come at the expense of reduced quality of tracking. The algorithms guarantee the robustness and accuracy of tracking as well as the extension of the overall system lifetime.

The rest of the chapter is organized as follows. The motivation is discussed in Section 6.1. Tracking applications are discussed in Section 6.2. The challenges associated with tracking using wireless sensor networks are presented in Section 6.3. In Section 6.4, an algorithm for tracking a dynamic phenomenon is addressed. In Section 6.5, general approaches to the tracking problem are presented. Numerical simulation results that validate the proposed algorithms are presented in Section 6.6 and the conclusions are summarized in Section 6.7.

6.1 Introduction and Motivation

Tracking, which involves identifying an object by its particular sensor signature and determining its path over a period of time, is one of the applications that can benefit from exploiting the characteristics of wireless sensor networks. The inherent parallelism of distributed sensors makes it possible to track multiple objects simultaneously, while the relatively low cost and ease of deployment enable the use of sensor network based tracking systems in remote or inaccessible locations, and when they need to be deployed on short notice. Algorithms for wireless sensor networks must have low communication overhead, rely as much as possible on local information, adapt to failures and changes in network conditions, and produce results in a timely fashion. Given the requirements to minimize the power, it is desirable to select the bare essential number of sensors dedicated for the task while all other sensors should preferably be in the hibernation or off state. Tracking mobile targets is an important application of sensor networks for both military and defense systems. Even though target tracking has been widely studied for sensor networks with large nodes and distributed tracking algorithms are available [50-61], tracking in ad hoc networks with micro sensors poses different challenges due to communication, processing and energy constraints. In particular, the sensors should collaborate and share data to exploit the benefits of sensor data fusion, but this should be done without sending data requests to and collecting data from all sensors, thus overloading the network and using up the energy supply. Target tracking is considered a canonical application for wireless sensor networks, and work in this area has been motivated in large part by DARPA programs.

In this chapter, results from the previous chapters are used to develop a distributed tracking algorithm using wireless sensor networks. As shown in the simulation results, the algorithm outperforms the other tracking algorithms in terms of tracking error and energy efficiency.

6.2 Tracking Applications

Research about object tracking in sensor networks varies from abstract models to projects designed to solve real life problems. Some papers studied object tracking in an abstract form and developed abstract models that either have been tested only by simulation, or tested on a specific application to serve as a proof of concept. Research in object tracking using sensor networks designed for a specific application spans a diverse variety of applications. One of the main applications for tracking using sensor networks is military surveillance and intruder detection. The Line in the Sand project [87] is a design of a sensor network used for military surveillance to track intruders and identify the intruder as being a vehicle, soldier or civilian. Simon et al. [88] presents a wireless sensor network design used to detect snipers and the trajectory of bullets. Another interesting application of object tracking using sensor networks is animal tracking and monitoring.

A wireless sensor network is being used to observe and track the behavior of zebras within a spacious habitat at the Mpala Research Center in Kenya [89]. Of particular interest is the behavior of individual animals (e.g., activity patterns of grazing, graze-walking, and fast moving, group behavior and group structure). The observation period is scheduled to last a year or more. The observation area may be as large as hundreds or even thousands of square kilometers. Butler et al. [90] presented a design of a sensor network to track a herd of cows and provide a virtual fence for the cows through collars holding sensor hardware that are attached to the cows' necks. In this case, the network did not only have to track cows but also take action to guide the cows to a certain point using acoustic microphones. Michahelles et al. [91] presents a design of a sensor used to assist rescue teams in saving people buried in avalanches. The goal of their work is to better locate buried people and to limit overall damage by giving the rescue team additional indications of the state of the victims and to automate the prioritization of victims (e.g., based on heart rate, respiration activity, and level of consciousness). The

GPS system is used in the avalanche rescue system to locate victims. The Smart Floor [92] presents a different application for tracking in sensor networks. The project aims at identifying people and tracking them around a room by using pressure sensors installed in a carpet. The Active Bat [93] system is an indoor tracking system that uses a matrix of ceiling mounted receivers used to track an object with a transmitter attached to it.

6.3 Performance Analysis and Challenges

Though certain types of energy harvesting are conceivable, energy efficiency will be a key goal for the foreseeable future. This requirement pervades all aspects of the system's design, and drives most of the other requirements.

In target tracking using wireless sensor networks, an important requirement of the sensor network is that the required data be disseminated to the proper end users. In some cases, there are fairly strict time requirements on this communication. For example, the detection of a poisonous gas intrusion in a surveillance network should be immediately communicated to the authority so that action can be taken. We describe the various issues associated with sensor networks that need to be addressed by any protocol being developed for application in sensor networks. We will outline some key design challenges for any proposed tracking algorithm in the domain of wireless sensor networks:

1- *Large number of sensors:* Networks of 10,000 or even 100,000 nodes are envisioned, so scalability is a major issue. Nodes may fail and new nodes may join the network. In the light of target tracking, the coordination function should scale with the size of the network, the number of targets to be tracked.

2- *Low energy use:* Since in many applications the sensor nodes will be placed in a remote area, service of a node may not be possible. In this case, the lifetime of a node may be determined by the battery life, thereby requiring the minimization of energy expenditure.

3- *Network self-organization:* Given the large number of nodes and their potential placement in hostile locations, it is essential that the network be able to self-organize; manual configuration is not feasible. Individual nodes may become disconnected from the rest of the network, but a high degree of connectivity must be maintained. Sensor nodes should be capable of organizing themselves into a network and achieving the desired objective in the absence of any human intervention or fixed patterns in the deployment.

4- *Collaborative signal processing:* The end goal is detection/estimation of some events of interest, and not just communications. To improve the detection/estimation performance, it is often quite useful to fuse data from multiple sensors.

5- *Distributed processing:* While a centralized architecture is theoretically optimal and also conceptually simple, it is not suitable in a large scale area because of the limited communication bandwidth of the wireless sensors. Moreover, failure of the fixed superior node may imply failure of the whole system.

6- *Tracking accuracy:* To be effective, the tracking system should be accurate and the likelihood of missing a target should be low.

7- *Computation and communication costs:* Any protocol being developed for sensor networks should keep in mind the costs associated with computations and communication. With current technology, the cost of computation locally is lower than that of communication in a power constrained scenario. As a consequence, emphasis should be put on minimizing the communication requirements.

8- *Uncertainty:* The exact positions of the nodes can not be known, so any position estimate of the target being tracked will be affected.

9- *Multi-modality sensor network:* The sensor have the abilities to sense the environment in various modalities , process the information , and forward it to a certain node for further processing. Compared to the single-modality sensor network that can only provide partial information of the environment, a multi-modality network can obtain more complete descriptions of the monitored environment through combining the fused data from various sensors with different capabilities and strengths.

10- *Time synchronization:* Time synchronization is a critical piece of infrastructure for any distributed system. Distributed, wireless sensor networks make particularly extensive use of synchronized time: for example, to integrate a time-series of proximity detections into a velocity estimate; to measure the time-of-flight of sound for localizing its source; to distribute a beamforming array; or to suppress redundant messages by recognizing that they describe duplicate detections of the same event by different sensors.

The impact of theses performance issues on the design of a tracking algorithm is addressed in the next sections. A distributed energy efficient tracking algorithm is presented. The algorithm specifically aims at minimizing the number of active nodes necessary to track a dynamic phenomenon while achieving a high level of tracking accuracy.

6.4 Distributed Tracking Algorithm

In real world, a sensor network is completely asleep for a long time. When some interesting event happens, only a limited zone of the network that is close to the event is kept in its fully active state. The active zone should be centered at the current location of a target phenomenon that is being tracked; and, of course, the zone should move through the network along with the target. Nodes that are not within sensing range of the event are outside of the zone, and therefore do not waste energy. Optimally, the zone should move such that a phenomenon of interest is always kept inside of the zone. The zone is a circular region where the center of this zone is the border sensor node which had detected the target. The radius of the zone depends on two factors: the maximum speed of the intruder and the maximum time needed to calculate a reduced cover. The key to the algorithm is that there is no central controller i.e. each node will decide autonomously to be active or not in order to track the target.

The sensor nodes in the network can be in three different modes:

1- Full Active Mode: A node is capable of both sensing and communicating with neighboring nodes.

2- Light Active Mode: A node can only communicate with neighbors.

3- Sleep Mode: A node is inactive.

The algorithms depend on the idea of "Divide and Conquer" which is basically selecting a reduced sensor cover for the region of interest, and if the phenomenon is moving, a new reduced sensor cover is established for the moving zone and so on. Every time the target is about to leave a zone, a border sensor node detects it and a new zone is created with the border sensor node as its center. A set of new sensor nodes within the circular zone and that belong to the reduced cover are activated. In order to save more energy, an enhancement to the algorithm would be adding prediction techniques where only a subset of the reduced cover nodes within the zone is activated depending on the predicted location of the target. Both approaches are discussed in details in the following sections.

Next, a number of algorithms that will aid us in developing a distributed tracking algorithm are provided.

Algorithm 1 *(Reduced Cover Algorithm)*

PROBLEM

Given a dense deployment of sensor nodes, find a minimum subset of active nodes that guarantee full coverage of **R**.

SOLUTION

The algorithm in chapter 3 indicate that a sensor node S_0 is completely covered if all the intersection points $C_i \cap C_j$ are covered by some sensor $S_l, l \neq i, j = 1..n$ where S_i and S_j are neighboring sensor nodes of S_0 and $C_i = A_0 \cap A_i$. Therefore, to check if S_0 is completely covered; one has to first find all the circles obtained by the intersection of $S_0 \cap S_k, k = 1..n$. For each C_k, find all the intersection points. If all these intersection points are covered by some sensor node, then the circles C_k are covered which implies that S_0 is covered and can be deactivated.

It is well known that the coverage problem in WSNs is NP-hard. The computational complexity of the algorithm developed in this section is $O(N^3)$ where $N = \left(\max_{i=1}^{n} |N(i)| \right)$ is the maximum number of nodes in the neighbor set of any sensor in the network.

193

Algorithm 2 *(Sub Reduced Cover Algorithm)*

PROBLEM

Given a reduced cover set of a region **R**, deduce the reduced cover of a sub region R_{sub} of **R**.

SOLUTION

Each node $S_j(x_j, y_j, z_j)$ that is part of the reduced cover set will receive an ALLERT message that contains the coordinates of the center of the sub region $S_i(x_i, y_i, z_i)$ and the maximum speed of the intruder. S_j will check if it lies within the sub region i.e. it will check if $d(S_i, S_j) \leq V_{max} . t_{max}$ and decides to be part of the reduced Sub Cover.

Algorithm 3 *(Boundary Cover Algorithm)*

PROBLEM

Given a reduced cover set of a region **R**, deduce the border cover.

SOLUTION

Each node $S_j(x_j, y_j, z_j)$ that is part of the reduced cover set will receive an ALLERT message that contains the coordinates of the center of the sub region $S_i(x_i, y_i, z_i)$ and the maximum speed of the intruder. S_j will check if its sensing region intersects the boundary

194

lines of the sub region i.e. it will check if $r_{sub} - R_s \le d(S_i, S_j) \le r_{sub} + R_s$ and decides to be part

of the reduced Sub Cover.

Algorithm 4 (*Prediction Algorithm*)

In the linear prediction (LP) model, also known as the autoregressive (AR) model, the

next location $X(n)$ is approximated by a linear combination of k past locations. We are

then looking for a vector 'a' of k coefficients, k being the order of the LP model.

Provided that the 'a' is estimated, the predicted value is computed simply by FIR filtering

of the k past samples with the coefficients using $X(n) = \sum_{i=1}^{k} a_i X(n-i)$. To keep the

calculation simple and the communication overhead low, the prediction model we use is

only based on the target's moving speed and its direction of movement using the previous

and current position of the target to predict the next location. The previous position of the

target $X(t-a) = (x(t-a), y(t-a))$ and the current position of the target $X(t) = (x(t), y(t))$ are

used to estimate the velocity and the direction of the movement. The velocity is given by

$v = \dfrac{d(X(t), X(t-a))}{a}$ while the direction is $\theta = \cos^{-1} \dfrac{x(t) - x(t-a)}{d(X(t), X(t-a))}$. The next position of the

target can be predicted by $x(t+a) = x(t) + vt \cos\theta$ and $y(t+a) = y(t) + vt \sin\theta$.

Tracking Algorithm

Energy efficient tracking of a target involves different steps:

Phase 1:

- Find a reduced cover of the region of interest.

- Deduce the border cover of the region of interest.

Phase 2:

- Detect the presence of the target.

- Broadcast the coordinates of the border sensor node and activate the necessary Sub reduced cover (Deduce the Sub border Cover).

- Move the sub region accordingly.

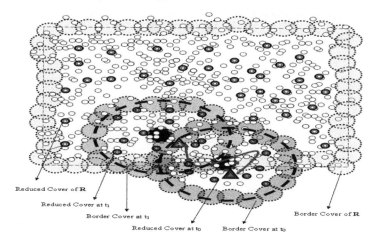

Figure 6.1: A snapshot of the distributed tracking algorithm in action.

The distributed tracking algorithm works by assigning a role for each sensor node. The initialization phase basically activates a border cover of the region of interest i.e. all the sensor nodes on the border of the region of interest are active. When a target is detected by a particular border sensor node S_i, then S_i is selected as the center node and broadcasts its coordinates to the reduced cover nodes in order to activate a subset of the reduced cover that will cover the sub circular zone of center S_i. The new sub border cover is deduced and as soon as a border sensor node detects the intruder, the same steps are repeated. This procedure guarantees the tracking of the target at all times since the radius of the circular zone depends on the maximum speed of the target and on the maximum time it takes to form the reduced cover of the sub region. A snapshot of the tracking algorithm is depicted in Figure 6.1. The flowchart of the processing performed at any given node S_i located at X_i to allow distributed target tracking is provided in Figure 6.2:

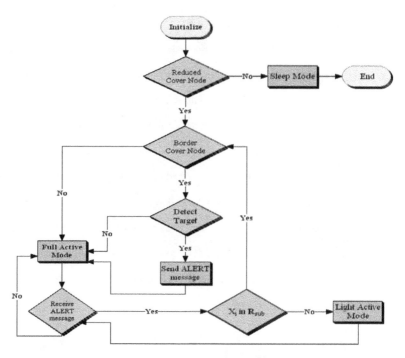

Figure 6.2: Flowchart of the processing performed at any given node using the tracking algorithm.

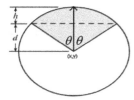

Figure 6.3: Circular sector of central angle 2θ.

A. Tracking Algorithm with prediction

Using the prediction algorithm, a sensor node estimates the targets next location moving with a velocity V and direction θ. Since the ultimate goal is conserve energy while achieving the necessary tracking performance, only a subset of the sensor nodes within a circle of radius R=V.T and center X is activated. Since the direction of motion is known, any sensor node that belongs to the reduced cover and is within a circular sector of central angel $\alpha = 2\theta$ is activated. So, instead of activating all the sensor nodes that lie within a circle of radius R and center X, we only activate a subset of theses sensor nodes that are with a circular sector of central angle 2θ where θ is the direction of the movement of the target (see Figure 6.3). The distributed tracking algorithm could be further optimized by encapsulating prediction techniques (Algorithm 4) and is performed using the following steps:

Step 1: Border Sensor node S_i detects target.

Step 2: S$_i$ broadcasts an ALERT message containing its location (x_i, y_i), the maximum velocity V_{max} , and the predicted direction of the moving target θ.

Step 3: Any sensor node S_j that receives the ALERT message, checks if it belongs to the reduced cover of the circular zone of center S_i. If so, it checks if the Euclidean distance between S_i and S_j is less than the radius of the tracking circular zone R. If that is the case, it checks if its within the circular sector of central angel 2θ. To do so, it checks if

ϕ, the angle between the straight line connecting S_i-S_j and the predicted direction of the target, is less than θ. If that is true, it decides to be active.

The flow chart of the processing at each sensor node using the predictive tracking algorithm is presented in Figure 6.4:

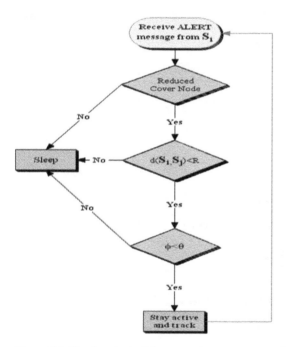

Figure 6.4: The flowchart of the processing at a sensor node using predictive reduced cover based tracking.

C. Performance Measures

We consider a sensor network consisting of n nodes deployed in some operational area, operating for a total time duration t. There is a single target moving through the area. We assume that all sensors in the network are binary detectors with a fixed sensing range R_s. In other words, at each instant, each sensor returns a '1' if the target is present within a distance S of that sensor, and a '0' otherwise. Given this simple sensor model, we take the centroid of the *locations* of all detecting sensors as an estimate of the target's location at any given time t_i. There are k sensors at locations $X_i(t_0) = (x_i(t_0), y_i(t_0))$ detecting the target at time t_0. Then, the estimated location of the target is:

$$X_{Target}(t_0) = (x_{Target}(t_0), y_{Target}(t_0)) \text{ where } x_{Target}(t_0) = \sum_{i=1}^{k} x_i(t_0), \ y_{Target}(t_0) = \sum_{i=1}^{k} y_i(t_0).$$

The two performance measures of interest to us in evaluating different tracking strategies are the coverage life time of the whole system (energy expenditure), and a measure of the tracking quality (accuracy), which reflects the uncertainty in the target's location. The two metrics are presented next.

Performance Measure 1 (*System Lifetime*)

We evaluate the system life time. The metrics used in evaluating system lifetime is the *coverage lifetime* of the region to be monitored. The overall coverage lifetime is the continuous operational time of the system before the coverage drops below its specified threshold (for example 0.9). Assuming that each sensor node has a limited energy supply

(300 Joules) and when it runs out of energy it is deactivated. The power consumption of Tx (transmit), Rx (receive), Idle and Sleeping modes are 1400mW, 1000mW, 830mW, 130mW respectively. As time passes, sensor nodes will be deactivated due to lack of energy and will leave some coverage holes in the border of the region.

Performance Measure 2 (*Tracking Error*)

We use the Euclidean distance between the estimated and actual locations of the target to measure the tracking error.

The tracking error at any given time t_i is:

$e(t_i) = d(X(t_i), X_{Target}(t_i))$ where $X(t_i)$ is the actual position of the target at time t_i and $X_{Target}(t_i)$ is the estimated position of target using the distributed tracking algorithm.

We denote the time average error over the total time t as $E = \frac{1}{t}\int_0^t e(t)dt$.

6.5 General Approaches to the Tracking Problem

Vlaam et al. [94] presents the design of an outdoor untagged sensor network system to detect multiple intruder location and velocity. Their system uses a centralized technique for communication and computation. However, their system suffers from serious limitations that make it highly impractical to use in a real application. First, the system requires that the nodes are distributed in such a way that the distance between the nodes is always the same. This assumption makes deployment less flexible. Second, the system uses passive infrared sensors which require line of sight and are not direction independent. Third, if an object is detected by a node and after a certain time period another object is detected by another node, it is assumed to be the same object because the passive infrared sensors cannot distinguish between two objects.

Some researchers have proposed distributed methods to overcome the problems of scalability and centralization in sensor networks by proposing distributed collaborative management techniques. The main motivation behind the idea of distributed management of object tracking sensor networks is that by limiting the collaboration to a small number of nodes in a limited geographical area, communication and computation load can be made independent of the size or area of the sensor network. Brooks et al. [54] propose a location-centric approach by dynamically dividing the sensor network into geographic cells run by a manager. Within each cell the manger node coordinates collaborative signal processing tasks. In the case of multiple measurements, they compare data fusion (combining data and then making a single decision) verses decision fusion (taking many local decisions and then combining them). Their approach can be summarized in five basic steps:

1- Nodes that have wireless cells near potential target trajectories are put on alert. Nodes within cells collaborate to determine if a target is present.

2- When a target is detected, the cell becomes active. If classification finds a target of the desired type, tracking is initiated.

3- Tracking includes estimating target location, direction, and speed for predicting future target positions. The target's next location is predicted using the last two actual locations of the target.

4- Based on the predictions, data from the active cell is sent to other cells; alerting them and facilitating Collaborative Signal Processing.

5- When the target is detected in an alerted cell, that cell becomes active and the process repeats.

Liu et al. [95] present a distributed method where collaborative sensor groups are formed, each responsible for tracking a single target. Their approach uses the same principles as [54]. However, they solve the problem of overlapping regions (or collision of tracks) when tracking multiple tracks by assigning a unique id to each track. The Line in the Sand project [87] exploits the fact that different metallic objects have different magnetic field signatures. The project defines the magnetic disturbance signature of a ferrous object to earth's magnetic field as the influence field. The project models the soldier or vehicle as a magnetic dipole, and then measures the minimum and maximum area in which it disrupts earth's ambient magnetic field. Their approach for tracking and identification can be summarized in the following steps:

1- A classifier passes a set or sets of nodes constituting each classified intruder and intruder type to the tracking module.

2- For the first time, the tracking module tags it as a new intruder.

3- It estimates the most likely intruder location as the centroid of the convex region enveloping all nodes detecting it.

4- It predicts an expected region for the intruder based on the velocity of the intruder type.

5- It correlates the tracked objects from successive windows in order to construct a continuous track per intruder.

6- If the expected and the estimated intruder region do not match, a new intruder is created.

Aslam et al. [96] proposes a method of using only one bit encoding to determine the direction information of the tracked object. The bits are broadcasted from the sensor nodes to a center station. This approach has a considerable impact on lowering the message size exchanged between the nodes. However, it requires extra sensors to provide proximity information if the location of the object, and not only the direction of the object, is to be determined. In their approach, the accuracy of the trajectory depends on the number of data points. Since their model has been only tested by simulation, it is not guaranteed that their model will work for a practical sensor network tracking application.

The issue of energy management in an object tracking sensor network has been gaining more interest recently. The work in [97 and 98] proposes an energy efficient technique using a sleep schedule where the nodes go to sleep when there is no need for sensing. Their algorithm makes use of the fact that the sensor network energy consumption is inversely proportional to the node density. Therefore, by reducing the coverage of non-critical areas we can save more energy switching those nodes in the non-critical areas to sleep mode. Gui et al. [99] uses the same principles for the algorithm in [97] but further divides the state of the sensor nodes into two states, surveillance state and tracking state.

The work in this book successfully combines all the strengths of other algorithm and at the same time eliminates their weaknesses. The algorithms are distributed, and minimize the number of sensor nodes active at one time while guaranteeing the quality of service needed. The system life time is extended and the delay of detection is minimized. In the next section, the distributed algorithm is compared to other popular distributed tracking algorithms and the advantages behind the work are highlighted.

6.6 Simulation Results

In this section, the theoretical results are validated through experimental simulations. The metrics developed in Section 6.4 are compared on different tracking strategies. The strategies that will be compared are:

1- Basic Strategy (BS): In this strategy, all the sensor nodes are in full active mode. Obviously this strategy offers the worst in terms of total coverage life time. However, it offers the best results in terms of tracking accuracy and thus serves as a baseline for comparison with other developed strategies.

2- Reduced Cover Strategy (RCS): In this strategy, the cover redundant sensor nodes are deactivated using algorithm 1 and all the remaining sensor nodes that are in the reduced cover track the object.

3- Basic Selective Predictive Strategy (BSPS: In this strategy, only a small subset of all the nodes is in tracking mode at any given point of time. They also predict the "next" position of the target and hand over tracking to nodes best placed to track the target in the "next" position. The rest of the nodes are in communication mode and can switch to tracking mode on being alerted by signals from tracking nodes. All the sensor nodes within a circle of specified radius centered on the predicted target location are in full active mode.

4- Reduced Cover Selective Tracking Strategy (RCSTS): In this strategy, only a subset of the reduced cover is activated to guarantee that the target is detected at all time.

5- Reduced Cover Selective Predictive Strategy (RCSPS): This strategy is an enhancement to RCSTS where only a subset of those activated using RCSTS are activated depending on the targets next predicted location.

We simulated a virtual large network of sensor nodes deployed on a 10x10 region of interest. A total of 1000 nodes were deployed. Different trajectories of the target were considered. First, the optimum coverage algorithm described in section 3 is used to find the reduced cover of region 10x10x10 units when sensor nodes are randomly deployed.

The nodes have a sensing radius of 1 unit and different numbers of nodes are randomly deployed in this region using a random distribution. It can be seen that the size of the reduced cover is almost the same as the number of deployed sensor nodes is increased, which indicates that the algorithms are scalable. Starting with 1000 deployed sensor nodes, the necessary number of nodes to be active in order to fully cover the region of interest is about 82 sensor nodes , resulting in a great deal of energy savings.

The performance of the network over time was also studied to determine the benefits of using a reduced cover tracking algorithm. This is done by assuming that each sensor node has a limited energy supply of 300 Joules and is deactivated when the available energy is used up. The performance is evaluated in terms of *coverage lifetime.* It can be seen in both the cases that the overall coverage drops over time as the available energy is used in processing the queries. Using the reduced network, it is seen that the resultant cover over time is significantly better. This is because each node in the reduced network has fewer neighbors and as a result has more efficient communications and less energy expenditure per query. This improvement in the coverage lifetime comes at a cost. The algorithm for obtaining the reduced network requires the communication between a node and its neighbors and as a result a portion of energy is used up during the initialization stage of the network. This causes early onset of degradation and loss of cover. This, however, can be addressed by incorporating *self healing* in the WSN.

Using the basic predictive strategy BSPS, there is significant improvement on the amount of energy consumed since only a subset of the nodes are active depending on the next location of the target. Comparison between the 3 algorithms is presented in Figure 6.5(a). Using the two proposed tracking algorithms (RCSTS and RCSPS), we notice that we can further improve on the energy savings and thus increasing the overall cover life time of the system as depicted in Figure 6.5(b). As we increase the number of deployed sensor nodes, the results are very similar however, we note the using BSPS decreases its performance since all the sensor nodes within a specified radius are activated while using the reduced cover approach, the number of sensor nodes to be activated would almost be the same since we first calculate the reduced cover of the region and then activate the necessary subset in order to track the target. The results are shown in Figure 6.5(c) and Figure 6.5(d).

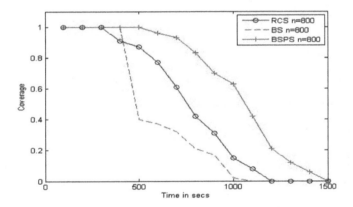

Figure 6.5(a): The coverage life time of the network as time passes using 3 different algorithms BS, RCS, BSPS when the number of deployed nodes is 800.

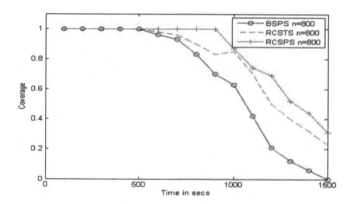

Figure 6.5(b): The coverage life time of the network as time passes using 3 different algorithms BSPS, RCSTS, RCSPS when the number of deployed nodes is 800.

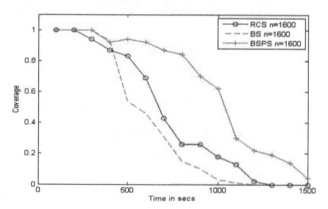

Figure 6.5(c): The coverage life time of the network as time passes using 3 different algorithms BS, RCS, BSPS when the number of deployed nodes is 1600.

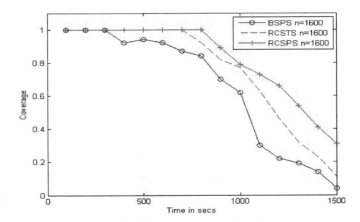

Figure 6.5(d): The coverage life time of the network as time passes using 3 different algorithms BSPS, RCSTS, RCSPS when the number of deployed nodes is 800.

The next experiment is more concerned with the accuracy of the tracking algorithm and compares the tracking error metric discussed in section 6.3 for the different tracking algorithms. We notice that in terms of tracking error, with no surprise, BS outperforms all the others. However, as the sensing radius of each sensor node is increased, all the other algorithms converge to a negligible tracking error. The results are shown in Figure 6.6.

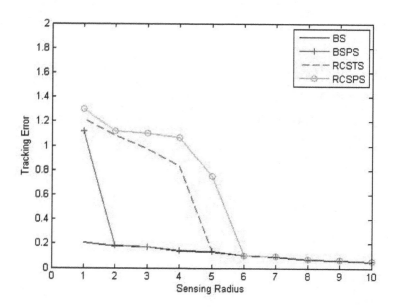

Figure 6.6: The resulting tracking error as we increase the sensing radius of each sensor node using 4 different algorithms: BS, BSPS, RCSTS, and RCSPS.

The simulation results show us the energy-quality trade offs between the different tracking algorithms. The algorithms minimize the number of active nodes while guaranteeing the moving target is detected at all times.

6.7 Chapter Conclusions

In this chapter, distributed tracking algorithms using wireless sensor networks were proposed. Theoretical as well as experimental results were developed. Unlike previous work in this area, the algorithms presented in this chapter make use of a minimal subset of sensor nodes in order to track a target. This minimizes the overall energy consumption and therefore, extends the lifetime of the network.

PART IV

CONCLUSIONS AND SCOPE FOR

FUTURE WORK

Chapter 7

Book Conclusions

In this book, the problems associated with the coverage of a wireless sensor network (WSN) and their impacts on the energy consumption were addressed. The limitations of existing techniques in the literature in determining the extent of coverage of the WSN when deployed in three dimensional spaces were studied. Computationally simple but elegant techniques were then developed for guaranteeing the coverage of the WSN using a minimal number of sensor nodes. The flexibility of the proposed approaches was also demonstrated for a range of applications requiring different types of coverage. The energy savings obtained using the proposed techniques was demonstrated through numerical simulations of the proposed techniques, as well as established techniques from the literature. The following sections highlight the contributions of the research presented in this book and the scope of the future work.

A. Technical Contributions

1. **Optimal Placement of Sensors.** The coverage problem in a three dimensional space was rigorously analyzed and the minimum number of sensor nodes and their placement for complete coverage was determined. This is a very important

215

result that will aid in the planning of wireless sensor networks. The numerical results in the book show that the total energy dissipation is a function of the number of active nodes and the ability to precisely place a minimum number of nodes will result in the most efficient deployment of a WSN.

The knowledge of the minimum number of nodes required for coverage is also a useful quantity in comparing different deployments of WSNs for coverage. The ratio of the total number of deployed nodes to the optimum number of sensor nodes defines a 'measure of optimality' that will enable the comparison of different implementations of a WSN from an energy efficiency stand point.

2. **Optimal Cover for Random Deployments of WSNs**. Many practical applications of WSNs do not allow the flexibility of placing the sensor nodes at optimum locations. The sensor nodes in these applications are randomly distributed, for example dropped from an airplane. In such cases, the problem of determining the coverage and selecting a minimum subset of sensor nodes for complete coverage is of paramount importance for the proper functioning of the network. In this book, a computationally efficient algorithm was developed that enables the deactivation of nodes that have overlapping coverage with neighboring nodes. This algorithm can be implemented in a distributed fashion and results in a measure of optimality close to '1'.

3. **Extent of Coverage and Border Coverage**. Practical deployments of WSNs do not always guarantee complete coverage of a specific region of interest. Therefore, it is necessary to determine the extent of the sensing cover and if any 'holes' exist in the coverage. The algorithms presented in Chapter 5, for the first time, provide algorithms that help determine the size and number of 'holes' in the coverage of deployments in the two and three dimensional regions.

4. **Efficient Border Cover and Border Perambulation**. Existing coverage algorithms do not provide the flexibility in choosing the sensor nodes depending on the needs of the application. The algorithms presented in Chapter 5 not only help determine the subset of nodes for covering the boundary of a given region, but also schedule the activation of nodes on the boundary for border perambulation. Application specific cover ensures that only the absolutely minimum number of sensor nodes is active at any given time, thereby ensuring minimum expenditure of energy and maximum lifetime of the WSN.

5. **Tracking of Dynamic Phenomenon**. One of the central issues in sensor networks is energy efficient target tracking, where the goal is to monitor the path of a moving target using a minimum subset of sensor nodes while meeting the specified quality of service (QoS). Unlike other tracking methods that are based on computationally complex clustering techniques, the strategy adopted in this book is based on a computationally simple but elegant technique of finding a reduced cover of the whole region and then subdividing the reduced cover into

sub-covers based on the target's location. The tradeoffs involved in target tracking are analyzed and the performance of the tracking algorithm is compared with other popular strategies from the literature. The behavior of the proposed tracking algorithm is analyzed through simulation and the improved performance is demonstrated. It is shown that there is a trade-off between the accuracy in tracking and the energy expended in the WSN. The gain in energy savings come at the expense of reduced quality of tracking.

6. **Self Healing for Fault Tolerant Operation**. The 'self healing' algorithms developed in this book ensure that the functionality of the WSN remains close to optimum even when the network is affected by faults or node outages. This property along with the optimization techniques proposed herein make possible the implementation of highly efficient, robust sensor networks whose performance is optimized with respect to the needs of the application.

The coverage algorithms developed in this book are a significant addition to the scientific knowledge in the area of wireless sensor networks. The proposed techniques help realize the practical deployment of wireless sensor networks in three dimensional regions. The algorithms presented can be easily extended to handle different shapes of region to be monitored. If we have region of an irregular shape, we can always use polygon approximation and simplification techniques to find the polygon that bounds the region of interest. In addition to that, the sensing radius of each sensor node need not be equal and the distributed algorithm could be applied to sensor networks with different sensing radii.

In this book, the sensing region of each sensor node was assumed to be a disk (2D) or an open ball (3D). This model is a simplification motivated by the need for analysis. This model, while useful, is still simplistic and does not address the realities in practical implementations. Signal to Noise ratio (SNR) in sensing, multi-path and fading issues in communications between the nodes, and data fusion are all important aspects that affect the functioning of a WSN and will be investigated in our future work. Our future work also evaluates the performance of our algorithms based on different sensing models and the design of hybrid coverage protocols capable of delivering accurate spatio-temporal profile of different kinds of sensing measurements.

We summarize the open research problems and point out the possible research directions in the area of sensor networks. The current research trends in object tracking using sensor networks as evident from recent published research papers are focusing on collaborative signal processing, distributed management of the network, energy efficient object tracking, and designing a sensor network to solve a specific tracking application. Some of the open research problems are summarized below.

B. Scope of the Future Work

1- **Localization and Topology Control in Sensor Networks:** The locations of sensor nodes are important for the meaning of collected data and for routing. The localization problem is to determine the node positions based only on the information obtained through the nodes' interactions, such as connectivity, edge lengths and angles. Topology control is the process of controlling the topology of a wireless network by adjusting the coverage ranges of its wireless nodes. Using rigorous combinatorial and probabilistic analysis, a future goal is to study various topological properties (e.g., connectivity, routing path, degree, local minimum for geographical routing, etc.), and present several variations of the topology control method. The findings should show balance between the various aspects of network performance.

2- **Power-aware System Design:** In recent years, there have been active developments of mechanisms that try to minimize power consumption while still providing some level of functionality in sensor networks. Many of these mechanisms operate at different levels. For example, new MAC layers try to optimize the link layer scheduling of multiple nodes to minimize power consumption. Topology control mechanisms try to exploit the redundancy in the communication networks to maximize network lifetime. Multiple routing mechanisms have been developed to minimize energy consumption. **Investigating the feasibility of integration of several of these mechanisms**

into one unique framework is crucial. Analytical studies and experimental evaluation to quantify the performance gains obtained by each particular scheme, and the energy effect of all the mechanisms operating at the same time should be preformed.

3- **Security in Wireless Sensor Networks:** Today, many networking and distributed systems are very vulnerable to faults or attacks, which can compromise the system performance, corrupt important data, or expose private information. Research on security has gained more and more attention and its major goal is to make systems more sustainable, secure and trustworthy. Future research will investigate security issues in networks and distributed systems, such as the resilience of peer-to-peer systems, the defense to distributed denial-of-service attacks, and privacy-preserving data mining. In contrast to traditional networks, wireless sensor networks encounter unique security problems due to their close interaction with physical environment.

4- **Object Tracking with Dynamic Sensor nodes:** All of the research that has been done assumes that the sensor nodes are static and are not mobile. The problem of tracking an object while the sensor themselves are mobile has not been researched yet. The idea of a mobile sensor network for tracking has not matured yet but applications such as ZebraNet indicate that such a network might exist or be needed in the future. A mobile sensor network for tracking

221

will require different (or extended versions) of current algorithms for communication, collaborative signal processing, and tracking.

5- **3D Tracking**: Applying the tracking problem to track systems in three dimensions, where the sensor network is three-dimensional and tracking needs to be in a three-dimensional matter. To the best of my knowledge, no one has researched the problem specifically or has indicated how to extend the two dimensional tracking problem to three dimensions.

6- **Data Mining in Smart Sensor Networks**: Sensors streaming their data online are turning the Internet into a global sensor network. Software platforms that integrate and mine these data streams may create a world in which sensors become pixels and we browse reality as easily as we browse Web pages today. The evolution of low cost, networked sensors, often directly Internet-enabled, is bringing sensors out of their traditional closed-loop realms into the rest of our reality. As sensor and communications technology continues to develop, we can envision a very different Internet than the one we use today. Rather than sending messages and browsing Web pages, we may experience new interactions such as experience sharing and browsing reality. Data mining, defined broadly as extracting useful information and insights from data, may be the untold half of the sensor networks story. Given the potentially huge amount of data streamed by live sensors, algorithms to fuse, interpret, augment, and present information will become an increasingly important part of networked sensor applications.

7- **Generalized Geographical Routing:** Geographical routing is a very important routing method, especially for ad hoc networks, mainly due to its high scalability. It is widely used for wireless networks, where it guides routing by using nodes' coordinates. Currently, the performance of geographical routing is limited by the hardness of the embedding process, the metric distortion caused by embedding, and its sensitivity to particular network models (such as the UDG model for wireless networks). A future goal would focus on studying new routing methods that generalize geographical routing, which use new backbone structures to guarantee message delivery and new shortcut links to guarantee efficiency.

8- **Time Synchronization in Wireless Sensor Networks:** Time synchronization is a critical piece of infrastructure in any distributed system. In sensor networks, a number of factors make flexible and robust time synchronization particularly important, while simultaneously making it more difficult to achieve than in traditional networks. Collaboration among nodes is often required for the data reduction that is critical to the energy efficiency of a sensor network. A future research goal would be to describe a spectrum of general design principles for sensor network time synchronization, and propose and implement a number of specific techniques.

References

[1]   M.Weiser, "The Computer for the 21st Century," Scientific American, pp. 94-104,
       September 1991.

[2]   M. Satyanarayanan, "Pervasive Computing: Vision and Challenges," IEEE Personal
       Communications, vol. 8, pp. 10-17, 2001.

[3]   D. Saha, A. Mukherjee, "A Paradigm for the 21st Century," IEEE Computer, IEEE
       Computer Society Press, vol. 36, pp. 25-31, 2003.

[4]   "Sensor Networks make Earlier Inroads," Extreme Sensor Networks, 2004.
       http://www.extremetech.com/.

[5]   X. Hong, M. Gerla, R. Bagrodia, P. Estabrook, T. Kwon, and G. Pei, " Load
       Balanced, Energy-aware Communications for Mars Sensor Networks," Proceedings
       of IEEE Aerospace Conference, Pasadena, CA, 2002, vol. 3, pp. 1109-1115.

[6]   NASA/JPL Sensor Webs Project, http://sensorwebs.jpl.nasa.gov/.

[7]   F. Akyildiz, D. Pompili, and T. Melodia, "Underwater Acoustic Sensor Networks:
       Research challenges," Ad Hoc Networks (Elsevier), vol. 3, pp. 257-279, 2005.

[8]   D. Pompili and T. Melodia, "Three-Dimensional Routing in Underwater Acoustic
       Sensor Networks," Proceedings of the 2nd ACM international workshop on
       Performance evaluation of wireless ad hoc, sensor, and ubiquitous networks,
       Montreal, Canada, 2005, pp. 214-221.

[9]   J.G. Proakis, E. M. Sozer, J. A. Rice, and M. Stojanovic, "Shallow Water Acoustic Networks," IEEE Communications Magazine, vol. 39, pp. 114-119, 2001.

[10] P. Gupta and P. R. Kumar, "The Capacity of Wireless Sensor Networks," IEEE Transactions on Information Theory, vol. 46, pp. 388-404, 2000.

[11] P. Gupta and P. R. Kumar, "Internet in the sky: The capacity of Three Dimensional Wireless Networks," Communications in Information and Systems, vol. 1, pp. 33-49, 2001.

[12] P. Mohapatra, "Mobile Ad Hoc and Sensor Networks," Technical Report, 2003. http://www.cs.ucdavis.edu/prasant/.

[13] R. Rammnathan, J. Redi, "A Brief Overview of Ad Hoc Networks: Challenges and Directions," IEEE Communications Magazine, 50th Anniversary Commemorative Issue, vol. 40, no. 5, pp. 20-22, 2002.

[14] J. Freebersyser, B. Leiner, "A DOD Perspective on Mobile Ad Hoc Networks," Ad Hoc Networking, ed. C, E. Perkins, Addison-Wesley, pp. 29-51, 2001.

[15] I. F. Akyildiz, et al., "Wireless Sensor Networks: A Survey," Computer Networks, vol. 38, no. 4, pp. 393-422, 2002.

[16] A. Galstyan, B. Krishnamachari, K. Lerman, S. Pattem, "Distributed Online Localization in Sensor Networks Using a Moving Target," Proceedings of the 3rd International Symposium on Information Processing in Sensor Networks, Berkeley, California, 2004, vol. 1, pp. 61-70.

[17] S. Meguerdichian, F. Koushanfar, M. Potkonjak, and M. Srivastava, "Coverage Problems in Wireless Ad-Hoc Sensor Networks," IEEE Infocom 2001, Anchorage, Alaska, 2001, vol. 3, pp. 1380-1387.

[18] S. Slijepcevic and M. Potkonjak, "Power Efficient Organization of Wireless Sensor Networks," Proceedings of IEEE International Conference on Communications, Helsinki, Finland, 2001, vol. 2, pp. 472-476.

[19] M. Cardei, D. MacCallum, X. Cheng, M. Min, X. Jia, D. Li, and D.-Z. Du, "Wireless Sensor Networks with Energy Efficient Organization", Journal of Interconnection Networks, vol. 3, no. 4, pp. 213-229, 2002.

[20] D. Tian and N. D. Georganas, "A Coverage-Preserving Node Scheduling Scheme for Large Wireless Sensor Networks," Proceedings of the 1st ACM Workshop on Wireless Sensor Networks and Applications, Atlanta, GA, 2002, pp. 32-41.

[21] F. Ye, G. Zhong, J. Cheng, S. Lu and L. Zhang, "PEAS: A Robust Energy Conserving Protocol for Long-lived Sensor Networks", International Conference on Distributed Computing Systems, Rhode Island, 2003, pp. 28-37.

[22] S. Kumar, T. H. Lai, and J. Balogh. "On k-coverage in a Mostly Sleeping Sensor Network," Proceedings of the 10th Annual International Conference on Mobile Computing and Networking (MobiCom 04), Philadelphia, PA, 2004, pp. 144-158.

[23] C. Liu, K. Wu, Y. Xiao, and B. Sun, "Random Coverage with Guaranteed Connectivity: Joint Scheduling for Wireless Sensor Networks," IEEE Transactions on Parallel and Distributed Systems, vol. 17, no. 6, pp. 562-575, 2006.

[24] H. Zhang and J. C. Hou, "Maintaining Sensing Coverage and Connectivity in Large Sensor Networks," Technical Report UIUC, UIUCDCS-R-2003-2351, June 2003.

[25] X. Wang, G. Xing, Y. Zhang, C. Lu, R. Pless, and C. D. Gill, "Integrated Coverage and Connectivity Configuration in Wireless Sensor Networks," First ACM Conference on Embedded Networked Sensor Systems (SenSys'03), Los Angeles, CA, 2003, pp. 28-39.

[26] K. Kar and S. Banerjee, "Node Placement for Connected Coverage in Sensor Networks," Proceedings of WiOpt 2003: Modeling and Optimization in Mobile, Ad Hoc and Wireless Networks, Sophia-Antipolis, France, 2003, pp.50-52.

[27] J. Wu and H. Li, "On Calculating Connected Dominating Set for Efficient Routing in Ad Hoc Wireless Networks," Proceedings of the 3rd International Workshop on Discrete Algorithms and Methods for Mobile Computing and Communications, Seattle, WA, 1999, pp. 7-14.

[28] K. Wu, Y. Gao, F. Li, and Y. Xiao. "Lightweight Deployment Aware Scheduling for Wireless Sensor Networks," ACM/Kluwer Mobile Networks and Applications (MONET), Special Issue on Energy Constraints and Lifetime Performance in Wireless Sensor Networks, 2004.

[29] M. Cardei and D. Du, "Improving Wireless Sensor Network Lifetime through Power Aware Organization", ACM Wireless Networks, vol. 11, no. 3, pp. 333-340, 2005 .

[30] S. Meguerdichian, F. Koushanfar, M. Potkonjak, and M. Srivastava, "Coverage Problems in Wireless Ad-Hoc Sensor Networks," IEEE Infocom 2001, Anchorage, Alaska, 2001, vol. 3, pp. 1380-1387.

[31] X.-Y. Li, P.-J. Wan, and O. Frieder, "Coverage in Wireless Ad-hoc Sensor Networks," IEEE Transactions on Computers, vol. 52, pp. 753-763, 2002.

[32] B. Liu and D. Towsley, "A Study of the Coverage of Large-scale Sensor Networks," Proceedings of the First IEEE International Conference on Mobile Ad hoc and Sensor Systems (MASS 2004), Fort Lauderdale, FL, 2004, pp. 475-483.

[33] S. Meguerdichian, F. Koushanfar, G. Qu, and M. Potkonjak "Exposure in Wireless Ad Hoc Sensor Networks," Proceedings of 7th Annual International Conference on Mobile Computing and Networking (MobiCom '01), Rome, Italy, 2001, pp. 139-150.

[34] S. Adlakha and M. Srivastava, "Critical Density Thresholds for Coverage in Wireless Sensor Networks", IEEE Wireless Communications and Networking, vol. 3, pp. 16-20, 2003.

[35] J. O'Rourke, "Art Gallery Theorems and Algorithms," Oxford University Press, Oxford, 1987.

[36] M. Marengoni, B. A. Draper, A. Hanson, and R. Sitaraman, "A System to Place Observers on a Polyhedral Terrain in Polynomial Time," Image and Vision Computing, vol. 18, pp. 773-780, 2000.

[37] W. W. Gregg, W. E. Esaias, G. C. Feldman, R. Frouin, S. B. Hooker, C. R. McClain, and R. H. Woodward, "Coverage Opportunities for Global Ocean Color in a Multimission Era," IEEE Transactions on Geosience and Remote Sensing, vol. 36, no. 5, pp. 1620-1627, 1998.

[38] D. W. Gage, "Command Control for Many-Robot Systems," Proceedings of the Nineteenth Annual AUVS Technical Symposium, AUVS-92, Hunstville, AL, 1992, pp. 22-24.

[39] K. Sohrabi, J. Pottie, "Protocols for Self-organization of a Wireless Sensor Network", IEEE Personal Communications, vol. 7, no. 5, pp. 16-27, 2000.

[40] C. Intanagonwiwat, R. Govindan, and D. Estrin, "Directed Diffusion: A Scalable and Robust Communication Paradigm for Sensor Networks," Proceedings of ACM MobiCom 2000, Boston, MA, 2000, pp. 56-67.

[41] F. Ye, A. Chen, S. Liu, L. Zhang, "A Scalable Solution to Minimum Cost Forwarding in Large Sensor Networks," Proceedings of the 10th International Conference on Computer Communications and Networks (ICCCN), Scottsdale, AZ, 2001, pp. 304-309.

[42] R. C. Shah and J. Rabaey, "Energy Aware Routing for Low Energy Ad Hoc Sensor Networks," IEEE Wireless Communications and Networking Conference (WCNC), Orlando, FL, 2002, vol.1, pp. 350-355.

[43] W. Heinzelman, A. Chandrakasan and H. Balakrishnan, "Energy-Efficient Communication Protocol for Wireless Microsensor Networks," Proceedings of the

33rd Hawaii International Conference on System Sciences (HICSS '00), Hawaii, 2000, pp. 3005-3014.

[44] Lindsey, S. and C. S. Raghavendra, "PEGASIS: Power-Efficient Gathering in Sensor Information Systems." Proceedings of IEEE Aerospace Conference, Pasadena, CA, 2002, vol. 3, pp. 1125–1130.

[45] A. Manjeshwar and D. P. Agarwal, "TEEN: A Routing Protocol for Enhanced Efficiency in Wireless Sensor Networks," In 1st International Workshop on Parallel and Distributed Computing Issues in Wireless Networks and Mobile Computing, San Francisco, CA, 2001, pp. 2009-20015.

[46] A. Manjeshwar and D. P. Agarwal, "APTEEN: A Hybrid Protocol for Efficient Routing and Comprehensive Information Retrieval in Wireless Sensor Networks," Proceedings of International Parallel and Distributed Processing Symposium (IPDPS), Fort Lauderdale, FL , 2002, pp. 195-202.

[47] B. Krishnamachari, D. Estrin, S. Wicker, "Modeling Data Centric Routing in Wireless Sensor Networks," Proceedings of the 21st Annual Joint Conference of the IEEE Computer and Communications Societies, New York, NY, 2002, pp.1-11 .

[48] Q. Li and J. Aslam and D. Rus, "Three Power-Aware Routing in Sensor Networks," Wireless Communications and Mobile Computing, vol. 2, no. 3, pp. 187-208, 2003.

[49] W. Heinzelman, J. Kulik, and H. Balakrishnan, "Adaptive Protocols for Information Dissemination in Wireless Sensor Networks," Proceedings of 5th ACM/IEEE MobiCom Conference (MobiCom '99), Seattle, WA, 1999, pp. 174-85.

[50] J. Kulik, W. R. Heinzelman, and H. Balakrishnan, "Negotiation-based Protocols for Disseminating Information in Wireless Sensor Networks," Wireless Networks, vol. 8, pp. 169-185, 2002.

[51] F. Zhao, J. Shin, and J. Reich, "Information-Driven Dynamic Sensor Collaboration for Tracking Applications," IEEE Signal Processing Magazine, vol. 19, pp. 61-72, 2002.

[52] F. Zhao, M. Chu, and H. Haussecker, "Scalable Information-Driven Sensor Querying and Routing for Ad hoc Heterogeneous Sensor Networks," International Journal of High Performance Computing Applications, vol. 16, no. 3, pp. 90-110, 2002.

[53] R. Brooks, C. Griffin, and D. Friedlander, "Self-Organized Distributed Sensor Network Entity Tracking," International Journal of High Performance Computer Applications, special issue on Sensor Networks, vol. 16, no. 3, pp.1-17, 2002.

[54] S. Phoha, N. Jacobson, D. Friedlander, and R. Brooks, "Sensor Network Based Localization and Target Tracking through Hybridization and Dynamic Space-time Clustering," Proceedings of IEEE 2003 Global Communications Conference (GlobeCOM), San Francisco, CA, 2003, pp. 1555–1567.

[55] R. Brooks, and C. Griffin, "Traffic Model Evaluation of Ad-hoc Target Tracking Algorithms," Journal of High Performance Computer Applications, vol. 16, no. 3, pp. 221-234, 2002.

[56] V. Cevher and J.H. McClellan, "Sensor Array Calibration via Tracking with the Extended Kalman Filter," Proceedings of IEEE International Conference on Acoustics, Speech, and Signal Processing (ICASSP), Salt Lake City, UT, 2001, vol. 5, pp. 2817-2820.

[57] J. Moore, T. Keiser, R. R. Brooks, S. Phoha, D. Friedlander, J. Koch, A. Reggio, and N. Jacobson, "Tracking Multiple Targets with Self-Organizing Distributed Ground Sensors," Journal of Parallel and Distributed Computing, vol. 64, no. 7, pp. 874-884, 2004.

[58] T. Clouqueur, V. Phipatanasuphorn, P. Ramanathan and K. Saluja, "Sensor Deployment Strategy for Target Detection," Proceedings of the 1st ACM International Workshop on Wireless Sensor Networks and Applications (WSNA'02), Atlanta, GA, 2002, pp. 42-48.

[59] K. Chakrabarty, S. S. Iyengar, H. Qi, E. Cho. "Grid Coverage of Surveillance and Target Location in Distributed Sensor Networks," IEEE Transaction on Computers, vol. 51, no. 12, pp. 1448-1453, 2002.

[60] B. Jung, and G.S. Sukhatme, "Tracking Targets using Multiple Robots: The Effect of Environment Occlusion," Autonomous Robots, vol. 13, no. 3, pp. 191-205, 2002.

[61] D. Li, K. Wong, Y. Hu and A. Sayeed, "Detection, Classification, Tracking of Targets in Micro-sensor Networks," IEEE Signal Processing Magazine, pp. 17-29, 2002.

[62] A. Cerpa and D. Estrin, "Ascent: Adaptive Self-configuring Sensor Networks Topologies," IEEE Transactions on Mobile Computing, vol. 3, no. 3, pp. 272-285, 2004.

[63] B. Krishnamachari, S. B. Wicker and R. Bejar, "Phase Transition Phenomena in Wireless Ad-hoc Networks," Proceedings of IEEE GLOBECOM, San Antonio, TX, 2001, pp. 2921-2925.

[64] E. Parker, B. Birch, and C. Reardon, "Indoor Target Intercept Using an Acoustic Sensor Network and Dual Wave Front Path Planning," Proceedings of IEEE International Symposium on Intelligent Robots and Systems (IROS '03), Las Vegas, NV, 2003, pp. 223-228 .

[65] L. M. Kaplan, Q. Le and P. Molnar, "Maximum Likelihood Methods for Bearings-only Target Localization," Proceedings of IEEE International Conference on Acoustic, Speech, and Signal Processing (ICASSP), Salt Lake City, UT, 2001, pp. 554- 557.

[66] J. Cortes, S. Martinez, T. Karatas, and F. Bullo, "Coverage Control for Mobile Sensing Networks: Variations on a Theme," Proceedings of IEEE ICRA, Lisbon, Portugal, 2002, pp.112-121.

[67] F. Zhao, J. Shin and J. Reich, "Information-driven Dynamic Sensor Collaboration for Target Tracking," IEEE Signal Processing Magazine, vol. 19, no. 2, pp. 61-72, 2002.

[68] J. Chang, K. Yao, T. Tung, C. Reed, and D. Chen" Source Localization and Tracking of a Wideband Source Using a Randomly Distributed Beamforming Sensor Array," International Journal of High Performance Computing Applications, vol. 16, no. 3, pp. 259-272, 2002.

[69] K. Yao, R. E. Hudson, C. W. Reed, T. L. Tung, D. Chen and J. C. Chen, "Estimation and Tracking of an Acoustic-seismic Source Using a Beamforming Array Based on Residual Minimizing Methods," Proceedings of IRIA-IRIS, Las Vegas, NV, 2001, pp. 153-163.

[70] L. Guibas, "Sensing, Tracking and Reasoning with Relations," IEEE Signal Processing Magazine, vol. 19, no. 2, pp. 73-85, 2002.

[71] Gregory J. Pottie, "Wireless Sensor Networks," in IEEE Information Theory Workshop, Santa Fe, New Mexico, 1998, pp. 139–140.

[72] S. Ni, Y. Tseng, Y. Chen, and J. Chen, "The Broadcast Storm Problem in a Mobile Ad-hoc Network," Proceedings of the Annual ACM/IEEE International Conference on Mobile Computing and Networking, Seattle, WA, 1999, pp. 151-162.

[73] H. Lim and C. Kim, "Multicast Tree Construction and Flooding in Wireless Ad-hoc Networks," Proceedings of ACM Modeling, Analysis, and Simulation of Wireless and Mobile Systems, Boston, MA, 2000, pp. 61-68.

[74] S. Basagni, I. Chlamtac, and D. Bruschi, "A Mobility-transparent Deterministic Broadcast Mechanism for Ad-hoc Networks," IEEE/ACM Transactions on Networking (TON), vol.7, pp.799-807, 1999.

[75] G. Fejes Toth, G. Kuperberg, and W. Kuperberg, "Highly Saturated Packings and Reduced Coverings," Monatsh Math., vol. 125, no. 2, pp.127-145, 1998.

[76] J. H. Conway and N. J. A. Sloane, "Sphere Packings, Lattices and Groups," Springer-Verlag, New York, 2nd edition, 1993.

[77] N. Megudio and A. Tamir, "On the Complexity of Locating Linear Facilities in the Plane," Operations Research Letters, pp. 194-197, 1982.

[78] F. Koushanfar, M. Potkonjok, and A. Sangiovanni, "Fault Tolerance Techniques for Wireless ad-hoc Sensor Networks," Proceedings of IEEE Sensors, vol. 2, pp. 1491-1496, 2002.

[79] Y. Zhang and K. Chakrabarty, "Adaptive Check pointing with Dynamic Voltage Scaling in Embedded Real-time Systems," Journal of Embedded Software for SoC , pp. 449-463, 2003.

[80] Y. Zhang, and K. Chakrabarty "Fault Recovery Based on Check Pointing for Hard Real-time Embedded Systems," Proceedings of the 18th IEEE International Symposium on Defect and Fault Tolerance in VLSI Systems, Boston, MA, 2003, pp. 320-327.

[81] K. Marzullo "Tolerating Failures of Continuous-valued Sensors," ACM Transactions on Computer Systems, vol. 8, no. 4, pp. 284–304, 1990.

[82] B. Krishnamachari and S. Iyengar "Efficient and Fault Tolerant Feature Extraction in Sensor Networks," Proceedings of the 2nd International Workshop on Information

Processing in Sensor Networks (IPSN '03), Palo Alto, CA, 2003, vol. 2634, pp. 488-501.

[83] Extreme Scale Wireless Sensor Networking - Technical report, http://www.cse.ohio-state.edu/exscal, 2004.

[84] R. Ghrist and A. Muhammad, "Coverage and Hole-detection in Sensor Networks via Homology," Proceedings of the 4th International Conference on Information Processing in Sensor Networks (IPSN'05), Los Angeles, CA, 2005, pp. 254-260.

[85] B. Carbunar, A. Grama and J. Vitek. "Distributed and Dynamic Voronoi Overlays for Coverage Detection and Distributed Hash Tables in Ad-hoc Networks," Proceedings of the Tenth International Conference on Parallel and Distributed Systems (ICPADS 2004), Newport Beach, CA, pp. 549-559, July 2004.

[86] R. Nowak, U. Mitra, "Boundary Estimation in Sensor Networks: Theory and Methods," Proceedings of the First International Workshop on Information Processing in Sensor Networks, Palo Alto, CA, 2003, pp. 80-95 .

[87] A. Arora, P. Dutta, S. Bapat, V. Kulathumani, H. Zhang, V. Naik, V. Mittal, H. Cao, M. G. M. Demirbas, Y. Choi, T. Herman, S. Kulkarni, U. Arumugam, M. Nesterenko, A. Vora, and M. Miyashita. "A Line in the Sand: A wireless Sensor Network for Target Detection, Classification, and Tracking," Computer Networks Journal, vol. 46, no. 5, pp. 605-634, 2004.

[88] G. Simon, M. Maroti, A. Ledeczi, G. Balogh, B. Kusy, A. Nadas, G. Pap, J. Sallai, and K. Frampton. "Sensor Network-based Counter Sniper System," Proceedings of

2nd International Conference on Embedded Network Systems, SenSys '04, Baltimore, Maryland, 2004, pp. 1-12.

[89] P. Juang, H.Oki, Y. Wang, M. Martonosi, L.-S. Peh and D. Rubenstein. "Energy-efficient Computing for Wildlife Tracking: Design Tradeoffs and Early Experience with ZebraNet," Proceedings of ASPLOS-X '02, San Jose, CA, 2002, pp. 96-107.

[90] Z. Butler, P. Croke, R. Peterson, and D. Rus. "Networked cows: Virtual Fences for Controlling Cows," Proceedings of WAMES, Boston, MA, 2004, pp. 221-227.

[91] F. Michahelles, P. Matter, A. Schmidt, and B. Schiele. "Applying Wearable Sensors to Avalanche Rescue." Computers and Graphics, vol. 27, no. 6, pp. 839-847, 2003.

[92] R. Orr and G. Abowd. "The Smart floor: A Mechanism for Natural User Identification and Tracking," Proceedings of Conference on Human Factors in Computing Systems, The Hague, Netherlands, 2000, pp. 126-139.

[93] M. Addlesee, R. Curwen, S. Hodges, J. Newman, P. Steggles, A. Ward, and A. Hopper, "Implementing a Sentient Computing System," IEEE Computer Magazine, vol. 34, no. 8, pp. 50-56, 2001.

[94] S. Vlaam. "Object Tracking in a Multi Sensor Network." Master's thesis, Deflt University of Technology, Computer Engineering Parallel and Distributed Systems, August 2004.

[95] J. Liu, J. Reich, P. Cheung, and F. Zhao. "Distributed Group Management for Track Initiation and Maintenance in Target Localization Applications," Proceedings of 2nd

International Workshop on Information Processing in Sensor Networks (IPSN), Palo Alto, CA, 2003, pp. 113-128.

[96] J. Aslam, Z. Butler, F. Constantin, V. Crespi, G. Cybenko, and D. Rus. "Tracking a Moving Object with a Binary Sensor Network," Proceedings of 1st International Conference on Embedded Network Systems, SenSys '03, Los Angeles, CA, 2003, pp. 150-161.

[97] T. Yan, T. Hee, and J. Stankovic. "Differentiated Surveillance for Sensor Networks," Proceedings of 1st International Conference on Embedded Network Systems, SenSys '03, Los Angeles, CA, 2003, pp. 51-62.

[98] T. He, S. Krishnamurthy, J. Stankovic, T. Abdelzaher, L. Luo, R. Storelu, T. Yan, L. Gu, J. Hui, and B. Krogh. "Energy Efficient Surveillance System using Wireless Sensor Networks," Proceedings of IMobiSYS'04, Boston, MA, 2004, pp. 270-283.

[99] C. Gui and P. Mohapatra. "Power Conservation and Quality Surveillance in Target Tracking Sensor Networks," Proceedings of MobiCom '04, Philadelphia, PA, 2004, pp. 129-143.

www.ingramcontent.com/pod-product-compliance
Lightning Source LLC
LaVergne TN
LVHW042332060326
832902LV00006B/121